Biocompatible l
Nanoparticles for Imaging of Cancer Cells

Debarati De

Table of Contents

List of abbreviations

AC	Alternating current
APTES	3-Aminopropyltriethoxysilane
BCA	Bicinchoninic acid
bcc	Body-centered cubic
CM	Cell membrane
CS	Chitosan
DA	Dopamine
DAPI	4,6-diamidino-2-phenylindole
DC	Direct current
DDS	Drug delivery systems
DMEM	Dulbecco's Modified Eagle's Medium
DMSO	Dimethyl sulfoxide
DNA	Deoxyribonucleic acid
DOX	Doxorubicin
DS	Critical diameter
EDAX	Energy dispersive analysis of X-rays
EDS	Energy dispersive spectroscopy
FA	Folic acid
FBS	Fetal bovine serum
fcc	Face centered cubic
FESEM	Field emission scanning electron microscope
FITC	Fluorescein isothiocyanate
FSC	Forward scatter
FWHM	Full width at half maximum
GAPDH	Glyceraldehydes-3-phosphate dehydrogenase
H_2DCFDA	2,7-dichlorodihydrofluorescein diacetate
HEPA	High-efficiency particulate air
HRTEM	High resolution transmission electron microscope

ITC	Isothermal Titration Calorimetry
MNPs	Magnetic nanoparticles
MRAM	Magnetic random access memory
MRI	Magnetic resonance imaging
MTT	3(4,5-dimethylthiazolyl-2) 2,5-diphenyl tetrazolium bromide assay
NMs	Nanomaterials
NPs	Nanoparticles
OA	Oleic acid
PBMC	Peripheral blood mononuclear cells
PBS	Phosphate buffered saline
PEG	Polyethylene glycol
PEI	Polyethyleneimine
PI	Propidium iodide
PMT	Photomultiplier tube
PVDF	Polyvinylidene difluoride
RCF	Relative centrifugal force
RES	Reticuloendothelial system
RITC	Rodamin B- isothiocyanate
ROS	Reactive oxygen species
RPMI	Roswell Park Memorial Institute
$RSiX_3$	Silane coupling agents
SAR	Specific absorption rate
SEM	Scanning electron microscope
SLP	Specific loss power
SQUID	Superconducting quantum interference device
SSC	Side scatter
TEM	Transmission electron microscope
TX-100	Triton X 100
UV-Vis	Ultraviolet-visible
VSM	Vibrating sample magnetometer
XRD	X-ray diffraction

List of symbols

A	Absorbance
B	Magnetic flux
C	Concentration
D	Dimensional
d	Crystal plane spacing
H	Magnetic field
h	Hour
Hc	Coercivity
I	Current
I	Intensity of light
K_a	Binding affinity
M	Magnetization
Mr	Remanence
Ms	Saturation magnetization
R	Gas constant
T	Temperature
t	Time
Tc	Curie temperature
V	Voltage
ε	Molar extinction coefficient
Θ	Angle
λ	Wave length
Φ	Magnetic flux
ΔG	Gibbs free energy changes
ΔH	Enthalpy changes
ΔS	Entropy changes

Chapter 1 Introduction

This introductory chapter narrates the literature review of nanoparticles, magnetic nanoparticles and their magnetic properties, crystal structures, anisotropy, hysteresis, coercivity, functionalization of the nanoparticles, targated drug delivery, magnetic hyperthermia, cellular uptake of nanoparticles and a brief explanation of these finding that creates a motivation to work in this specific research field.

Recently, magnetic nanoparticles (MNPs) with their tunable size, shape, and porosity have become a subject of immense interest in the area of research due to its important applications in biomedical sciences. The magnetic property of the MNPs can be controlled by varying their structure, composition, size which opens up the chance of various applications such as magnetic separation, magnetic resonance imaging (MRI), magnetic hyperthermia, magnetic field triggered drug delivery and release etc. Moreover, the MNPs also own chemical stability, uniform dispersion in liquid medium, controlled size and easy surface modifications. Among the MNPs, both superparamagnetic and ferromagnetic nanoparticles are compatible for magnetic hyperthermia therapy of cancer treatment because of their production of local heat under AC magnetic field. Among the different cancer treatments, the magnetic hyperthermia has special advantage because of their less adverse side effects. Beside these new types of proposed treatments, some time for use of cancer drug if a new technique can be developed for specific local use of drugs that will also be very much beneficial for cancer patients as the cancer drugs are very harmful and they have adverse side effects on human body. As the cancer drug is equally harmful for both normal and cancer cell, hence the use of this drug for treating of cancer without affecting normal cell is a great challenge. But researchers have overcome this problem by applying targeted drug delivery that minimize the side effects and also increase the effectiveness of drug. Targeted drug delivery involves different types of carrier molecules to which drugs become loaded. Among the carriers the MNPs takes special interest because of their unique magnetic property. Drug loaded MNPs can be moved to the desired location by using an external magnetic field. In the MNPs mediated drug delivery, the release of drug can also be controlled by magnetically induced heat treatment methodology due to their property to induce heat under AC magnetic field.

In this thesis, we have discussed about the synthesis of different types of magnetic nanoparticles with variable size, shape and magnetic property. We have modified their surface with different organic molecules to increase their biocompatibility and stability in solution. The physico-chemical properties of these MNPs have been observed by using different techniques such as XRD, TEM, SEM, FTIR, ITC and the magnetic properties has observed by SQUID and VSM. For biomedical applications such as magnetic hyperthermia and drug delivery, different cancer cells have been cultured at 37°C in 5% CO2 incubator and details cellular experiments have been done. Here, we have tuned the magnetic property of the MNPs and check their heating ability under AC magnetic field. We have also tried to increase the drug loading and releasing efficiency under different stimuli. For cellular imaging, we have invented a new dye from beet root and checked their fluorescence imaging on different cancer and normal cell lines.

Introduction:

1.1 Nanoparticles:

Nanoparticles (NPs) are wide class of materials including particulate substances, which have at least one dimension less than 100 nm ($1 nm=10^{-9}$ m) [1,2]. Nanoparticles have variety of shapes such as nanospheres, nanorods, nanofibers, nanoflowers, nanochains, nanostars, nanoreefs, nanoboxes, nanowhiskers etc [3-6]. Depending on the overall shape, these materials are classified as 0 Dimensional (D), 1D, 2D or 3D [7]. Researchers found that some applications of nanoparticles may require specific shapes and sizes as they can influence the physiochemical properties of a substance. Due to different properties of the surface from the core the NPs are considered to be composed of three different layers i.e. the surface layer, the shell layer, and the core. Usually, the surface layer consists of a variety of small molecules such as metal ions, surfactants and polymers. On other hand, the shell layer is chemically different material from the core when the core refers the nanoparticle itself [8]. A NP may contain with a single material or a combination of several materials. Depending on chemical and electromagnetic properties, usually nanoparticles can exist as suspensions, colloids or dispersed aerosols. Based on morphology, size and chemical properties, the NPs are classified into various categories such as carbon based nanoparticles [9,10], metal NPs [11],ceramic NPs [12,13],polymeric NPs [14,15], semiconductor NPs [16,17],lipid based NPs [18,19] etc. Nanoparticles have variety of applications in different fields such as in medicine [20,21], catalysis [22-24], energy storage [25,26], sensing [27], information technology [28], environment [29,30]etc. Among different types of nanoparticles, we have used magnetic nanoparticles for our experimental work because of their some properties such as they are cheap, easy to prepare, stable and have suitable optical and magnetic properties that can also be manipulated easily by changing of their size, shape and composition.

1.2 Magnetic nanoparticles:

Magnetic nanoparticles (MNPs) are a class of nanoparticles which can be operated by magnetic field. In general, such materials are made of iron, cobalt, nickel etc. or their oxides with unpaired'd' electrons, which have key role in the development of magnetic properties. Due to their small size they show some unique properties which is different from their bulk and therefore they have potentiality in wide range of applications such as catalysis [31,32], magnetic recording, magnetic random access memory (MRAM), magnetic fluids [33], magnetic switches [34,35], contrast agents in magnetic resonance imaging (MRI) [36], water treatment [37], biotechnology [38], bio imaging [39], bio separation [40], protein separation [41], biomedicine [42], organ repair [43], drug delivery [44,45], gene delivery [46], drug carrier [47], cancer cell destruction [48], targeted therapy [49,50], hyperthermia [51,52] etc. For different biomedical applications, the MNPs with superparamagnetic behaviour at room temperature are preferable [53-55]. In addition, for applications in biological and medical fields, the MNPs should also be stable in water at physiological condition i.e. at pH7. The colloidal stability of the MNPs in solution depends on surface charges of the MNPs which causes both steric and coulombic repulsions and also depend on the particles dimension which should be sufficiently small that precipitation can be avoided by the gravitation forces [56]. The magnetic nanoparticles can also be used for in vivo applications for which the particles must be biocompatible. Therefore, particles encapsulated or coated with a biocompatible material are very potential candidates. Sometimes, polymer and the other organic materials coating help adsorption or covalent attachment of different drugs with the nanoparticles [57-59]. For in vivo applications, the MNPs must be made of a non-toxic and non-immunogenic material and the particles size should be small enough which remain in the circulation inside the body and can easily pass through the capillary of organs and tissues. The MNPs must also have good magnetization that their movement inside the body can be controlled with a magnetic field which helps in targeting therapy [60-62]. The synthesis method of nanomaterials determines the shapes, size distribution, size, the surface chemistry and magnetic properties of the particles [63-65]. The MNPs show some new phenomena i.e. high saturation field, high field irreversibility, superparamagnetism, extra anisotropy contributios, shifted loops after field cooling etc. These phenomena originate from the size (narrow and finite) and surface effects which dominate the magnetic behaviour of the individual MNPs [66]. According to Frenkel and Dorfman [67], a ferromagnetic material whose particle size is below 15 nm would consist of a single magnetic domain and the particle shows uniform magnetization at any field.

1.3 Crystal structure of MNPs:

The magnetic nanoparticles with spinel structures have a chemical formula MFe_2O_4 where M can be different divalent transition metal ions (Mn^{2+}, Fe^{2+}, Co^{2+}, Ni^{2+}, Zn^{2+} etc.). The spinel ferrites crystallize in the face centered cubic (fcc) spinel structure which belongs to space group Fd-3m. The spinel configuration consists of a cubic close-packed structure with 56 atoms where 32 of them are oxygen anions and rest of the 24 cationic metal ions occupy 8 of the 64 available tetrahedral sites (A sites) and 16 of the 32 available octahedral sites (B sites) represents in Fig.1.1. Generally in normal spinel structures, the tetrahedral sites contain the 8 M^{2+} ions while the 16 Fe^{3+} ions all occupy the octahedral sites express by a formula:$[M^{2+}]^A[Fe^{3+}]_2{}^BO_4$. But in case of inverse spinel configurations, all the 8 M^{2+} ions occupy the 8 octahedral sites and 16 Fe^{3+} ions are divided into 8 tetrahedral and 8 octahedral sites represents by formula: $[Fe^{3+}]^A[M^{2+}Fe^{3+}]^BO_4$ [68]. The structural formula for a typical spinel ferrites can be written as $[M^{2+}{}_{(1-x)}Fe^{3+}{}_x]^A[M^{2+}{}_xFe^{3+}{}_{(2-x)}]^BO_4$ while x corresponds to the inversion degree [69].

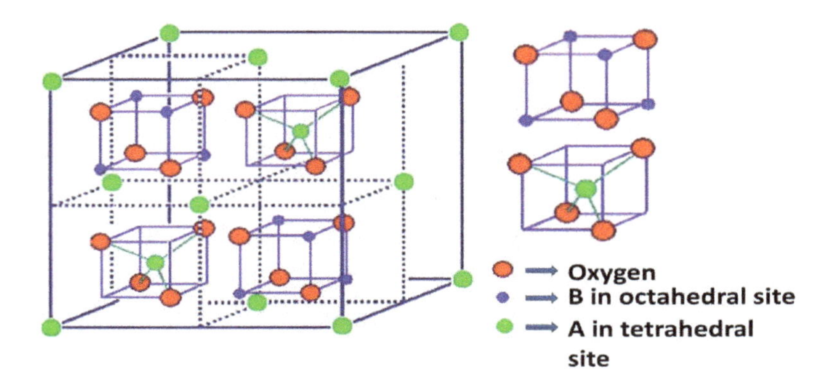

Fig.1.1. Crystal structure of typical spinel ferrites.

1.4 Magnetic properties of MNPs:

Generally, the magnetic effects of MNPs are created by movements of electrons, holes, protons and also by positive & negative ions present in the particles. Spinning of an electrically charged particle creates a magnetic dipole, also called magneton. In ferromagnetic materials, magnetons are associated in groups and in a magnetic domain; all magnetons are aligned in the same direction by the exchange forces. This domain concept differentiates ferromagnetism from paramagnetism. The domain structure of a ferromagnetic material depends on the size of the particles which determines its magnetic behaviour. A ferromagnetic material becomes a single domain when its size is reduced below a critical value. The ferromagnetic particles with smaller size have uniform magnetization when in the larger particles, the non-uniform magnetization has been observed. Therefore, according to particles size, the ferromagnetic materials are referred to as single domain particles and multidomain particles [70,71]. Following the magnetic domain theory, several factors affect the critical size of the single domain such as the value of the magnetic saturation, strength of the crystal anisotropy, surface or domain wall energy, the shape of the particles etc. A ferromagnetic material shows some reaction under an applied field which is described by a hysteresis loop, characterized by two main parameters such as remanence (M_R) and coercivity (H_C). The coercivity is strongly size dependent and observed that as the particle size is waned, the coercivity increases to a maximum and then decreases toward zero as shown in Fig. 1.2. In case of superparamagnetic materials, the size of single domain particles further decreases below a critical diameter and the coercivity becomes zero. Generally, in the presence of an external magnet, the nanoparticles become magnetic but the NPs also revert to its nonmagnetic state when the external magnet is removed. This behaviour of NPs gives a unique advantage to use of these particles in biological purposes. There are a number of MNPs such as oxides of iron (Fe), cobalt (Co), manganese (Mn), and nickel (Ni) which show good magnetic, optical, electronic & catalysis properties, high chemical stability, nontoxicity, cost effectiveness etc. [72-75]. The nanoparticles (NPs) which have high saturation magnetization (M_S) and low coercivity (H_C) are preferable for bio-sensing, bio-separation due to their high sensitivity and efficiency when NPs with higher H_C are required for information storage. Hence, before application, magnetic property of the NPs must be tuned according to their uses.

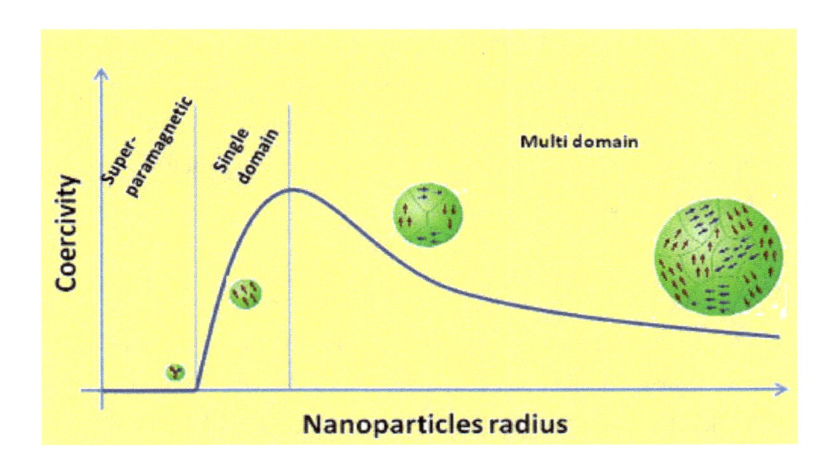

Fig. 1.2. Schematic representation showing relation of coercivity with particle size.

1.5 Magnetic anisotropy:

Magnetic anisotropy describes about a material's magnetic properties which can be diffentiate depending on growth of the particles in different directions. Some factors such as crystal structure, shape of the materials etc. have an influence on the direction of magnetization. The shape of hysteresis loops are strongly affected by the magnetic anisotropy and it also controls the coercivity (H_C) and remanence (M_R). There are different types of anisotropy depending on several parameters: magneto-crystalline anisotropy, shape anisotropy, and magnetoelastic anisotropy.

1.5.1 Magneto-crystalline anisotropy:

Magneto-crystalline anisotropy is an intrinsic property of a ferrimagnet which is independent of grain size and shape. Magneto-crystalline anisotropy is the energy which is needed to change the magnetic moment in a single crystal from the easy to the hard direction of magnetization. The easy and hard directions of a magnetic sample originate from the interaction of the spin magnetic moment with the crystal lattice known as spin-orbit coupling. This potent spin-orbit coupling combines the orbital motion of electrons with the crystal electric field and also makes the contiguous spins to be parallel or anti-parallel to each other. The magnitudes of the magneto-crystalline anisotropy can be monitor by the ratio of the crystal field energy and spin orbit coupling [76]. The magnetic anisotropy can describe by two models: Néel model and single-ion or crystal field model. According to Néel model, the magnetic anisotropy arises between two magnetic ions because of their inter-ionic interactions [77] where single-ion or crystal field model describes the interaction of crystal field with the magnetic atoms [78]. The magneto-crystalline anisotropy energy can be exposed by the general formula:

$$E_k = V(K_0 + K_1 \sin^2\theta + K_2 \sin^4\theta)$$

where, Kn ($n = 0, 1, 2 \ldots$) denote the magneto-crystalline anisotropy constants and θ console the angle between magnetization and easy direction of the crystal. In cubic structure of magnetite particles, the magneto-crystalline anisotropy energy can be expressed by a series expansion of the angles between the magnetization direction and the cube axes. Generally, at higher temperature, the magneto-crystalline anisotropy decreases with thermal agitation.

1.5.2 Shape anisotropy:

Shape anisotropy depends on the grain shape and its magnitude is dependent on the saturation magnetization. In magnetite particles, the shape anisotropy is the dominant form of anisotropy due to its high saturation magnetization. But in case of hematite materials, the shape anisotropy is usually never important because of its very low saturation magnetization. For larger sized particles, they become multidomain with good crystalline structure. Therefore, in case of larger sized particles, the shape anisotropy is generally less important than the magneto-crystalline anisotropy.

1.5.3 Magneto-elastic anisotropy:

In addition to magneto-crystalline anisotropy, there is another anisotropy that related to spin-orbit coupling known as magneto-elastic anisotropy where the strain induces stress in thin magnetic films through the spin-orbit interaction [79]. In magneto-elastic anisotropy, the spin moments are coupled to the lattice via the orbital electrons. Whether the lattice is changed by strain, the distance within the magnetic atoms is altered. Therefore, the interaction energy is changed [80]. The magnitude of the magneto-elastic anisotropy depends on the level of stress.

1.6 Magnetic domains:

A magnetic material consists of some region called magnetic domains where groups of magnetic moments align in the same direction. The magnetic behaviour of ferromagnetic materials (iron, nickel and cobalt) and ferrimagnetic materials (ferrite) is due to their domain structures. Under an external magnetic field, the domains of a magnetic material are rotate and align with the external magnetic field and when almost all domains are aligned in the same direction, the whole material becomes magnetized. The magnetic domains are separated by regions called domain walls where the magnetization changes its direction from one domain to another domain. French physicist Pierre-Ernest Weiss first suggest about the existence of magnetic domains in ferromagnets [81] Depending on grain size, the magnetic behaviour of the magnetic particles can be divided into three ranges [82]. Multidomain particles have larger size whose magnetization reversal process involves the nucleation and motion of domains. When the grain size of the particles decreases, the grain becomes a single domain.

1.7 Magnetic hysteresis and magnetic coercivity:

A magnetic hysteresis also known as hysteresis loop is a plot represents the variation of magnetization (M) with the strength of magnetic field (H) of a ferromagnetic material. Generally, two phenomena affect the hysteresis of ferromagnetic materials i.e. rotation of magnetization and the changes of the size or number of magnetic domains. The magnetization of a magnet mainly varies with its direction but not depends on its magnitude. In case of single domain magnets, the magnetization is produced by rotation under a magnetic field. In general, magnets consist of domains and the domain walls. The walls can move with the changes of magnetic field which causes changing of the relative sizes of the domains. Initially, the domains of a magnet are magnetized in one direction which needs some extra energy to demagnetize again. The coercivity (H_C) of a magnetic material is measured from the magnetic hysteresis loop. The H_C of a ferromagnetic material defined as the intensity of the applied magnetic field (H) which is required to demagnetize that material, after reaching the particle's magnetization to saturation. To evaluate the coercivity, the H field is required to reduce the magnetic flux (B), magnetization (M) and remanence (M_R) to zero.

1.8 Surface modification of MNPs:

The MNPs have a tendency to aggregate together because of their large surface to volume ratio and high surface energies. Furthermore, the bare MNPs are highly chemically active and get oxidized easily in air [83]. Due to this reasons, the surface modification of MNPs is very necessary. Besides the increase of stability of MNPs via preventing agglomerations and oxidations, surface modification also helps in further functionalization of the particles [84]. Previous study reports that the aspartic-acid absorbed MNP shows good colloidal stability at an increased pH [85]. Sometimes, prior functionalization of NP by some special chemical groups helps in proper surface modification of these particles for particular use. Such groups are –OH, -COOH,-NH_2,-SH etc. There are wide varieties of molecules which have been used to modify the surface of the MNPs are divided into such groups such as inorganic small molecules, organic small molecules, organic polymers etc. as shown in Fig. 1.3. Among the inorganic molecules the SiO_2 takes an important attention to modify the MNPs. SiO_2 is an acidic oxide which has some properties such as stable chemical properties, biocompatibility, low toxicity, acid resistance and high temperature resistance etc. SiO_2 is rich in hydroxyl groups [86] which make covalent bonds between the MNPs and proteins, DNA, antibodies & some bioactive molecules [87]. Some noble metals such as gold (Au) and silver (Ag) are preferable to used for surface modification of MNPs due to their novel properties like chemically stable, biocompatible, resistant to oxidation and corrosion etc [88]. Au modification increases the stability of MNPs under physiological conditions and also enhances their binding ability to ligand. Beside this, Au can also prevent the formation of harmful free radicals [89]. Zhao et al. found that the aptamers functionalized Au-Fe_3O_4 acts as a good drug carrier which is highly sensitive to diagnostics and therapy of diseases [90]. According to Chen et al., Ag modification on the surface of Fe_3O_4 NPs shows good antibacterial property to Staphylococcus and Escherichia coli [91]. In addition these inorganic molecules, some organic molecules basically surfactants also play an important role in surface modifications. Oleic acid (OA), a surfactant is nontoxic in blood and highly soluble in organic solvents (ethanol, chloroform, ether). OA increases the biocompatibility, dispersibility, stability of MNPs and also reduced the surface energy [92]. The carboxyl group of OA make a strong chemical bond with the surface of the MNP which increases the dispersibility [93]. Another surfactant citric acid is also acts like OA, having three carboxylic acid groups which can be covalently bound with the iron ions of the MNPs. The electrostatic repulsive forces of citrate prevent the aggregation of nanoparticles and increase the despersibility. The citric acid modified MNPs have different applications for example it can be used for in vivo stem cells tracking [94]. Moreover, some other surfactants are also be used to modify the surface of MNPs such as glutamic acid, alicylic acid, trichloroacetic acid which prevent the nanoparticles aggregation in the biological environment [95]. These surfactants have some free groups after being bound to MNPs that can help in further functionalization. Some silane coupling agents ($RSiX_3$) are used for MNPs surface modification. Here, R represents an active group, and X represents an alkoxy group. The silane-coupling agent contains organic and inorganic groups which make bond with MNPs and exhibit both the hydrophilic and lipophilic properties. 3-Aminopropyltriethoxysilane (APTES), an amino silane coupling agent shows well-dispersibility, superparamagnetic behaviour, and strong magnetization at room temperature after binding it with MNPs [96]. Other silane coupling agent such as vinylsilane and methacryloxysilane are also used as a surface modified agents [97,98]. There are some organic polymers used to modify MNPs such as chitosan (CS), dextran, polyethyleneimine (PEI), polyethylene glycol (PEG) etc. The CS has many suitable properties like biocompatibility, biodegradability, nontoxicity and hydrophilicity. The CS has used in the wide variety of fields including health care, food hygiene [99] and waste treatment [100]. Dextran coating increases the biocompatibility of MNPs and also prevents the aggregation [101]. Moreover, the PEI-modified MNPs have several applications in targeted drug delivery, magnetic transfection, magnetic hyperthermia etc [102]. PEG is a hydrophilic, nontoxic, long chain polymer which has potential applications in diagnosis and therapy of cancer [103]. Some surface active groups of MNPs can pair with DNA, protein or antibody which broadens the particles application scopes [104].

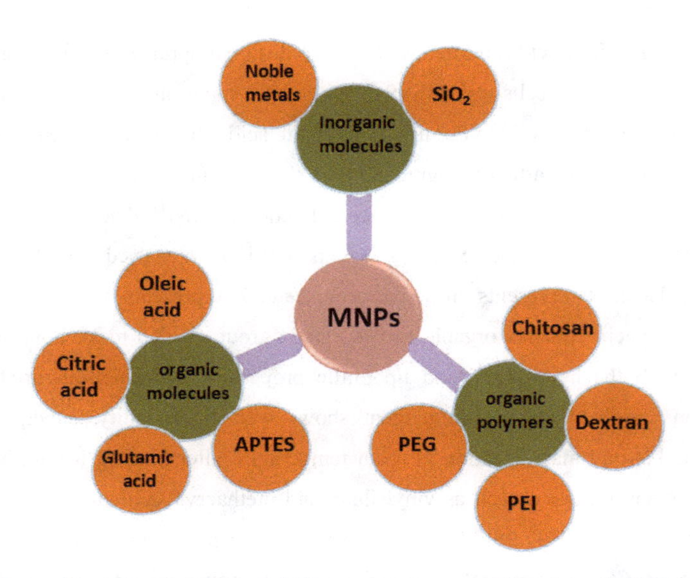

Fig. 1.3. Schematic representation of different surface modification of MNPs.

1.9 Targeted drug delivery:

The drug delivery system is one kind of process in which drug is administered orally or injected intravenously and finally drug will be spread throughout the body via blood circulation. By this system, small quantity of drugs can enter into the target tissue and also side effects may occur where the healthy tissues being affected. To overcome these problems, targeted drug delivery is very effective method where the drugs are delivered selectively to the targeted organs through some special techniques. In these techniques, the efficacy of drugs is increased significantly by reducing their side effects [105]. The drug delivery by using magnetic nanopaticles (MNPs) known as magnetically targeted drug delivery is a very hopeful application of in biomedical science [106,107]. In MNPs mediated drug delivery, the drugs are injected into the body after attaching them with MNPs. Here, the circulation of the drugs loaded MNPs can be controlled by an external magnetic field which facilitates the accumulation of drugs to the diseased site. Inside the body, the NPs below 10 nm can be easily cleared by a renal route [108], for which they are not able to play a role in the target organ. Therefore, the particle sizes ranged between 10 to 100 nm are optimal for drug delivery due to their longer blood circulation time [109]. In addition, the particle shape can also play a role in blood circulating time. Scientist has noticed that anisotropic MNPs are more efficient to circulate in the blood compare to spherical shaped MNPs [110]. Surface modification is always necessary to stabilize the MNPs in the aqueous physiological environment. Thus, modification of the MNPs with some organic and inorganic molecules including PEG, dextran, chitosan, phospholipids, polyethyleneimine and silica is essential [111-116]. Another important requirement for in vivo drug delivery is to reduce opsonisation of the MNPs by the immune cells, which is usually prevented by coating the MNPs with protein repellent molecules such as PEG [117]. The final step of intracellular delivery drug loaded MNPs is closely dependent on the particle size, adhesion to cell membrane, endocytosis efficiency of cells etc [118]. In magnetically assisted drug delivery, the most important aspect is that the controlled delivery of drugs to target site under external magnetic fields. This method based on the binding of drugs to the MNPs and after reaching the target site, desorption of drug from the MNPs by external magnetic field. Certain factors influence the magnetic targeted drug delivery system such as strength of the applied magnetic field, magnetic properties of MNPs, drug loading capacity, particles size, blood flow rate etc.

1.10 Magnetic hyperthermia:

Recently, hyperthermia has received newish attention due to their application in cancer therapy. In magnetic hyperthermia, the MNPs have been used as heat mediators for novel thermotherapy treatments of cancer. MNPs are one of the ferromagnetic materials which can produce heat under an alternating magnetic field [119,120]. The hyperthermia treatments for cancer cells mainly based on heat generation within the cancer cells. Generally, tumour cells have higher sensitivity to temperature compared to normal cells [121,122]. In this treatment, the MNPs are subjected into the tumour site and focused to generate the heat inside the tumour in the range of 41°C to 47°C under an alternating magnetic field [123]. The heat treatment causes cancer cell death by changing of their cellular growth, differentiation, protein structures and enzymes etc at this specified temperature [124]. The heating efficiency of the MNPs basically affected by their size and magnetic susceptibility [125]. This method usually does not damage the normal tissues. Several studies show that the tumour size has been reduced after combination of hyperthermia treatment with other therapies like radiotherapy and chemotherapy [126-128].

1.11 Cellular uptake of nanoparticles:

Cellular uptake of nanoparticles involves the mechanism where the NPs make complex biomolecular interactions to enter through the cell membrane (CM). The cell membrane, also called plasma membrane encloses the cytoplasm by separating the cell's interior fluid from the exterior fluid. CM plays some important role such as protects the intracellular components, gives the cell structure, maintains the cell homeostasis, retains the cell composition etc [129-133]. The CM mainly consists of phospholipids, proteins and some biomolecules that results overall negative charge of the membrane with some cationic domains. The CM is selectively permeable to ions, biomolecules, nanoaprticles etc. For cellular entry, the NPs need to overcome the cell membrane. Types of uptake pathway determine the NP's function, intracellular fate and biological response [134-136]. During in vivo and in vitro applications, the nanoparticles enter through the CM by following different routes: (i) endocytosis-based cellular entry and (ii) direct cellular entry represented in Fig. 1.4. There are five distinct classes of endocytosis based entry pathway including (a) clathrin-dependent endocytosis; (b) caveolin-dependent endocytosis; (c) clathrin- and caveolin-independent endocytosis; (d) phagocytosis; and (e) macropinocytosis. After entring inside the cell, the NPs are enclosed within the intracellular vesicles like endosomes, phagosomes, macropinosomes etc. Among the endocytosis process, phagocytosis is an uptake process where the any external foreign bodies (pathogens, diseased cells, synthetic/biological materials) are engulf and cleared by immune cells like macrophages, dendritic cells, neutrophils, and B lymphocytes [137]. Phagocytes recognize the NPs and clear from the circulation with high efficiency [138,139]. Nanoparticle's phagocytosis occurs by physical binding of NP to the surface receptors of phagocyte cell. Molecular mechanism of recognition and clearance of foreign substances by the immune cells is known as opsonization. To reduce the opsonization, surface modification of nanoparticles is required [140]. One of the important molecule used for this surface coating is poly(ethylene glycol) (PEG) [141]. PEG density and its degree of polymerization on the surface of NP affect the NP opsonization and blood circulation time [142].

Direct entry of NPs into the cells also involves different paths: (a) cytoplasmic entry by direct translocation; (b) cytoplasmic entry by lipid fusion; (c) electroporation and (d) microinjection. In case of direct entry, NPs can cross the cell membrane directly and enters into the cytoplasm. For direct cellular entry, the NPs must be very small in size (<5 nm). Researchers found that cellular uptake of NP are affected by the cell type [143]. The cellular uptake also depends on the size of nanoparticles [144]. In addition, both the shape anisotropy and orientation of NPs control its uptake in cell [145,146]. Surface charge plays an important role in the interaction with cells and found that positively charged NPs are more efficiently internalized into the cell compared to negatively charged and neutral NPs even at similar dimensions [147].

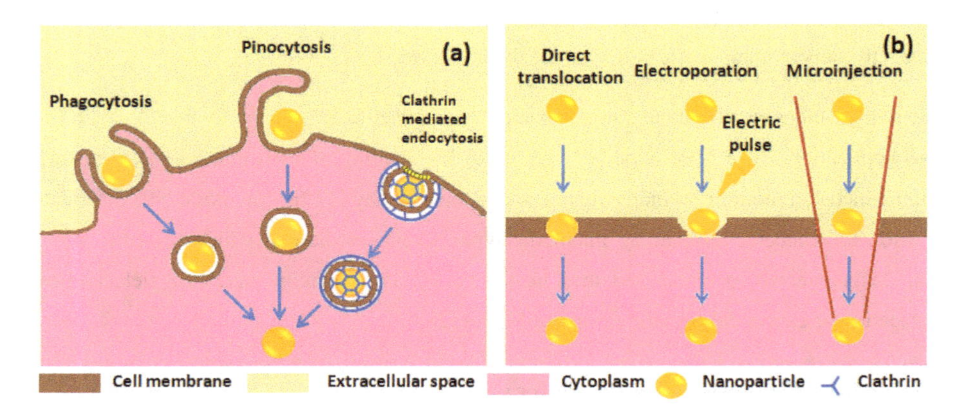

Fig. 1.4. Different routs of cellular entry of NPs (a) endocytosis-based cellular entry and (b) direct cellular entry.

1.12 Iron oxide nanoparticles:

Iron is a chemical element with symbol Fe which is the fourth most common element in the earth's crust. Iron exists in several allotropic forms such as α-iron, β-iron, γ-iron, and δ-iron. The α-iron is magnetic and stable to 768° C. The crystal structure of this type of iron is body-centered cubic (bcc). β-iron is stable between 768° C to 910° C which is mainly alpha iron that has lost its magnetism. γ-iron is nonmagnetic and stable between 910° C to 1390° C, crystallizes in face-centered cubic (fcc) where the δ-iron is nonmagnetic and stable between 1391° C and the melting point is approximate 1539° C. The oxidation form of iron i.e. iron oxides are consists of iron and oxygen. There are sixteen iron oxides, divided into three main groups: Iron (II) Oxide (wustite, FeO), mixed iron (II) and (III) Oxide (magnetite, Fe_3O_4) & Iron (III) Oxide (Hematite, Fe_2O_3). Among these magnetite have attracted due to their superparamagnetic properties and their potent applications in different fields. Magnetite has an inverse spinel structure with fcc crystal structure. In magnetite materials, all tetrahedral sites are occupied by Fe^{3+} where octahedral sites are occupied by both Fe^{3+} and Fe^{2+}. An iron atom has a strong magnetic moment due to its four unpaired electrons in 3d shell ([Ar] $3d^6$ $4s^2$). Iron oxides in its Fe^{2+} state have also four unpaired electrons in 3d shell where Fe^{3+} have 5 unpaired electrons in 3d shell. Magnetite is a ferrimagnetic material at room temperature with a Curie temperature (T_C) of 850 K. Ferromagnetic materials lose their magnetization beyond the Curie temperature. Under an external magnetic field, the magnetization of a ferrimagnetic material increases with the strength of the magnetic field until approaches saturation. The magnetite NPs are superparamagnetic at room temperature [148]. Particle size affects the superparamagnetic behaviour of the iron oxide NPs.

Due to their biocompatibility and non- toxicity, the magnetite NPs are preferable to use in in-vivo applications. The iron oxide NPs are used for labelling of cells (stem cells, dendritic cells) which can monitor by magnetic resonance tomography [149]. Iron oxide NPs may also be used for cancer treatment by magnetic hyperthermia method in which the iron oxide solution is injected to tumor and then heated up by an alternating magnetic field causes destruction of cancer cells inside the tumor [150-152].

1.13 Cobalt ferrite nanoparticles:

Among various spinel ferrites, cobalt ferrite ($CoFe_2O_4$) nanoparticles (NPs) are one of the valuable MNPs. All or most Co^{+2} ions of $CoFe_2O_4$ NPs occupied the octahedral sites (B sites) where the Fe+3 ions occupied on both tetrahedral (A sites) and B sites [153,154]. The $CoFe_2O_4$ NPs show high anisotropy, high coercivity at room temperature, good saturation magnetization (80emu/g) [155-157]. In addition, the $CoFe_2O_4$ NPs possess worthy physicochemical properties, excellent stability and colloidal dispersibility under physiological conditions. Because of these features, researchers have been working on the use of this nanoparticles in sensors, solar cells, recording devices, magnetic cards,bioseparation and purification, biotechnology, magnetic drug delivery, imaging and magnetic hyperthermia [158-163]. The specific absorption rate (SAR) of the $CoFe_2O_4$ NPs is increased with increasing the amount of Co ions [164]. Therefore, the magnetic properties of $CoFe_2O_4$ NPs are dependent on the Co content of the NPs which determines the heating ability of the NPs. The higher anisotropy of Co ions may be improves the heating ability of the NPs.

1.14 Motivation and objectives of thesis:

Nanoparticles take an important position in research area because of its many novel properties which are different from their bulk. In case of biomedical applications, researchers chose NPs due to its tiny size which can easily enter and removed from the body. Among the different types of nanoparticles, the ferrite based NPs have several possibility of application in the biomedical field as well as technological field. The ferrite NPs can be used for drug delivery, bio separation, hyperthermia, information storage, magnetic sensors and others because of their suitable properties such as low cost, chemical and mechanical stability, moderate magnetic saturation, high magneto-crystalline anisotropy, high electrical resistance, high phase-transition temperature and low thermal conductivity [165,166]. On other side, we know that cancer is the second leading cause of death in the world and the survey report that approximately 70% deaths from cancer occur in poor and moderately poor countries. There are several treatments for cancer such as radiation therapy, chemotherapy, surgery etc. But all these treatments can not cure cancer properly and these are also very much painful for patients and costly. During the last few decades, the researchers try to use the MNPs for cancer treatment by applying magnetic hyperthermia and drug delivery [167,106]. We are motivated from these works to use of the ferrite MNPs in hyperthermia and drug delivery for cancer treatment. Throughout our research work, we have focused on several motives such as drug loading and releasing efficiency, targeted delivery to cancer cells, different stimuli dependent drug release, decrease or prevention of the toxicity of drug loaded particles to healthy cells, increase of the biocompatibility; circulation time, better imaging of cancer cells, better coercivity, low cost etc. There are many parameters which can change the magnetic properties of the MNPs. For example, NPs show greater coercivity in spherical shape than the cubic although of their same volume, when the saturation magnetization and remanence remain unchanged [168]. The coercivity (H_C) of the NPs is significantly controlled by the particles size which increases with decrease of the particle size but there is a critical diameter, below which the H_C decreases to zero. Proper functionalization can enhance the H_C of the NPs along with its size and morphology [169]. Surface modification of the NPs increases their biocompatibility and circulation time.

In this frame the main objectives of this thesis are summarized as follows:

- Synthesis of micelle guided cobalt ferrite ($CoFe_2O_4$) nanoaprticles (NPs) by chemical co-precipitation method and studies their details DC and AC magnetic properties for possible hyperthermia application with drug delivery under influence of AC magnetic field.
- Synthesis of micelle free $CoFe_2O_4$ NPs and DNA functionalized $CoFe_2O_4$ NPs by chemical co-precipitation method and study of their AC magnetic property for hyperthermia application on breast cancer cell lines along with different in-vitro cellular studies.

- Synthesis of cube shaped magnetite (Fe_3O_4) nanoparticles by chemical co-precipitation method to study their AC magnetic properties and in-vitro drug release rate under influence of different stimuli such as pH and temperature for cancer treatment.
- Surface modification of drug loaded $CoFe_2O_4$ NPs by polyethylene glycol to make the particles more stable in blood circulation for their use as drug carrier for lung cancer treatment and studied in detail about cellular changes after drug delivery.
- Imaging of cancer cells by tagging the NPs with RITC dye. Modification of the nanoparticles surface with organic molecules for better imaging of cells.
- Extraction of organic dye from natural sources to use them as fluorescent marker for visualizing the different cancer cells.

1.15 Organization of the thesis:

The entire thesis is organized into five different chapters. A summary of the chapters is given below:

Chapter 1 begins with an introduction where we have discussed about magnetic nanoparticles, their crystal structures and magnetic properties. Here we also give information regarding the necessity of surface modification of MNPs for different applications. Moreover, we describe the hyperthermia and drug delivery method, cellular uptake procedure of NPs and also about iron oxide and cobalt ferrite NPs. Furthermore, the motivation behind the thesis work and the summery of the work has been discussed in this chapter.

Chapter 2 discusses the synthesis procedure of different magnetic nanoparticles and their various characterization techniques with instrumental details.

Chapter 3 demonstrates the synthesis of micelle coated cobalt ferrite NPs and their detail characterization including their DC and AC magnetic properties for use of these particles for possible hyperthermia applications. This chapter also discuss about the synthesis of $CoFe_2O_4$ NPs and DNA functionalized $CoFe_2O_4$ NPs including their characterization and biomedical application. We have checked the cellular entry of these MNPs to breast cancer cell lines (MDAMB-231) by fluorescence tagging and MTT assay was performed to check in-vitro cytotoxicity effect of the MNPs. AC magnetic measurements of these MNPs were done and hyperthermia applications were carried out. Moreover, we also discuss about the cell viability rate and the cellular changes which occur due to hyperthermia therapy. Furthermore, this chapter discuss about the extraction of organic dye from beet-root and labelling of the

$CoFe_2O_4$ NPs with this dye. Different cancer cell imaging studies were done with the raw dye and the dye labelled $CoFe_2O_4$ NPs.

Chapter 4 discusses the different synthesis procedures of MNPs (Fe_3O_4&$CoFe_2O_4$), characterization, drug loading, surface modification and use of these materials for drug delivery. Drug release rate from the MNPs under different stimuli (pH& temperature) has been observed by UV-Vis absorption spectroscopy. Furthermore, we also performed details cellular experiments and discuss the cellular toxicity, internalization of particles, apoptotic pathways, cellular changes that occur after treatment with drug loaded $CoFe_2O_4$ NPs.

Chapter 5 concludes this thesis and proposes about the scope of future work in this related research area.

Bibliography:

[1]. Laurent, S., Forge, D., Port, M., Roch, A., Robic, C., Elst, L.V., Muller, R.N., 2010. Magnetic iron oxide nanoparticles: synthesis, stabilization, vectorization, physicochemical characterizations, and biological applications. Chem. Rev., 110, 2574–2574.

[2]. Vert, M., Doi, Y., Hellwich, K. H., Hess, M., Hodge, P., Kubisa, P., Rinaudo, M., Schué, F. O., 2012. Terminology for biorelated polymers and applications (IUPAC Recommendations 2012). Pure and Applied Chemistry, 84 (2), 377-410.

[3]. Agam, M. A., Guo, Q., 2007. Electron Beam Modification of Polymer Nanospheres. Journal of Nanoscience and Nanotechnology, 7 (10), 3615–9.

[4]. Kralj, S., Makovec, D., 2015. Magnetic Assembly of Superparamagnetic Iron Oxide Nanoparticle Clusters into Nanochains and Nanobundles. ACS Nano, 9 (10), 9700-9707.

[5]. Choy, J.H., Jang, E.S., Won, J.H., Chung, J.H., Jang, D.J., Kim, Y.W., 2004. Hydrothermal route to ZnO nanocoral reefs and nanofibers. Appl. Phys. Lett., 84 (2), 287-289.

[6]. Sun, Y., Xia, Y., 2002. Shape-controlled synthesis of gold and silver nanoparticles. Science, 298 (5601), 2176-2179.

[7]. Tiwari, J.N., Tiwari, R.N., Kim, K.S., 2012. Zero-dimensional, onedimensional, two-dimensional and three-dimensional nanostructured materials for advanced electrochemical energy devices. Prog. Mater Sci., 57, 724–803.

[8]. Shin, W.-K., Cho, J., Kannan, A.G., Lee, Y.-S., Kim, D.-W., 2016. Cross-linked composite gel polymer electrolyte using mesoporous methacrylate-functionalized SiO2 nanoparticles for lithium-ion polymer batteries. Sci. Rep.6, 26332.

[9]. Astefanei, A., Nu´ n˜ ez, O., Galceran,M.T., 2015. Characterisation and determination of fullerenes: a critical review. Anal. Chim. Acta, 882,1–21.

[10]. Ibrahim, K.S., 2013. Carbon nanotubes-properties and applications: a review. Carbon Lett., 14, 131–144.

[11]. Dreaden, E.C., Alkilany, A.M., Huang, X., Murphy, C.J., El-Sayed, M.A., 2012. The golden age: gold nanoparticles for biomedicine. Chem. Soc. Rev., 41, 2740–2779.

[12]. Sigmund, W., Yuh, J., Park, H., Maneeratana, V., Pyrgiotakis, G., Daga, A., Taylor, J., Nino, J.C., 2006. Processing and structure relationships in electrospinning of ceramic fiber systems. J. Am. Ceram. Soc., 89, 395–407.

[13]. Thomas, S., Harshita, B.S.P., Mishra, P., Talegaonkar, S., 2015.Ceramic nanoparticles: fabrication methods and applications in drug delivery. Curr. Pharm. Des., 21, 6165–6188.

[14]. Mansha, M., Khan, I., Ullah, N., Qurashi, A., 2017. Synthesis, characterization and visible-light-driven photoelectrochemical hydrogen evolution reaction of carbazole-containing conjugated polymers. Int. J. Hydrogen Energy, XXX, 1-10.

[15]. Rao, J.P., Geckeler, K.E., 2011. Polymer nanoparticles: preparation techniques and size-control parameters. Prog. Polym. Sci., 36, 887–913.

[16]. Ali, S., Khan, I., Khan, S.A., Sohail, M., Ahmed, R., Rehman, A.,Ur Ansari, M.S., Morsy, M.A., 2017. Electrocatalytic performance of Ni@Pt core–shell nanoparticles supported on carbon nanotubes for methanol oxidation reaction. J. Electroanal. Chem., 795, 17–25.

[17]. Khan, I., Abdalla, A., Qurashi, A., 2017a. Synthesis of hierarchical WO3 and Bi2O3/WO3 nanocomposite for solar-driven water splitting applications. Int. J. Hydrogen Energy, 42, 3431–3439.

[18]. Puri, A., Loomis, K., Smith, B., Lee, J.-H., Yavlovich, A., Heldman, E., Blumenthal, R., 2009. Lipid-based nanoparticles as pharmaceutical drug carriers: from concepts to clinic. Crit. Rev. Ther. Drug Carrier Syst., 26, 523–580.

[19]. Gujrati, M., Malamas, A., Shin, T., Jin, E., Sun, Y., Lu, Z.-R., 2014. Multifunctional cationic lipid-based nanoparticles facilitate endosomal escape and reduction-triggered cytosolic siRNA release. Mol. Pharm., 11, 2734–2744.

[20]. Veiseh, O., Tang, B.C., Whitehead, K.A., Anderson, D.G., Langer, R., 2014. Managing diabetes with nanomedicine: challenges and opportunities. Nat. Rev. Drug Discov., 14 (1), 45-57.

[21]. Juliano, R., 2013. Nanomedicine- is the wave cresting? Nat. Rev. Drug Discov., 12 (3), 171–172.

[22]. Garadkar, K.M., Ghule, L.A., Sapnar, K.B., Dhole, S.D., 2013. A facile synthesis of ZnWO4 nanoparticles by microwave assisted technique and its application in photocatalysis. Mater. Res. Bull., 48 (3), 1105–1109.

[23]. Grunwaldt, J.D., Kiener, C., Wögerbauer, C., Baiker, A., 1999. Preparation of Supported Gold Catalysts for Low-Temperature CO Oxidation via "Size-Controlled" Gold Colloids. J. Catal., 181, 223–232.

[24]. Hashmi, A.S.K., Hutchings, G.J., 2006. Gold Catalysis. Angew. Chemie Int. Ed., 45 (47), 7896–7936.

[25]. Aricò, A.S., Bruce, P., Scrosati, B., Tarascon, J-M., Schalkwijk W.V., 2005. Nanostructured materials for advanced energy conversion and storage devices. Nat. Mater., 4, 366-377.

[26]. Larcher, D., Tarascon, J-M., 2014. Towards greener and more sustainable batteries for electrical energy storage. Nat. Chem., 7, 19-29.

[27]. Zhang, W., Li, X., Zou, R., Wu, H., Shi, H., Yu, S., Liu, Y., 2015. Multifunctional glucose biosensors from Fe3O4 nanoparticles modified chitosan/graphene nanocomposites. Sci. Rep., 5, 11129.

[28]. Corr, S.A., 2013. Metal oxide nanoparticles. Nanoscience: Volume 1: Nanostructures through Chemistry, 1, 180–207.

[29]. Ripp, S., Henry, T.B. (Eds.), 2011. Biotechnology and Nanotechnology Risk Assessment: Minding and Managing the Potential Threats around Us. ACS Symposium Series. American Chemical Society, Washington, DC, DC., 1079.

[30]. Zhuang, J., Gentry, R.W., 2011. Environmental application and risks of nanotechnology: a balanced view. ACS Symposium Series; American Chemical Society: Washington, DC, 41–67.

[31] Sun, S., 2006. Recent Advances in Chemical Synthesis, Self-Assembly, and Applications of FePt Nanoparticles. Adv. Mater, 18, 393–403.

[32] Cao, S-W., Zhu, Y-J., Ma, M-Y., Li, L., Zhang, L., 2008. Hierarchically Nanostructured Magnetic Hollow Spheres of Fe3O4 and γ-Fe2O3: Preparation and Potential Application in Drug Delivery. J. Phys. Chem., C 112, 1851–1856.

[33] Rabani, E., Reichman, D.R., Geissler, P.L., Brus, L.E., 2003. Drying-mediated self-assembly of nanoparticles. Nature, 426, 271-274.

[34] Martín, J.I., Nogués, J., Liu, K., Vicent, J.L., Schuller, I.K., 2003. Ordered magnetic nanostructures: fabrication and properties. J. Magn. Magn. Mater, 256, 449–501.

[35] Cox, A.J., Louderback, J.G., Apsel, S.E., Bloomfield, L.A., 1994. Magnetism in 4d-transition metal clusters. Phys. Rev.B, 49, 12295–12298.

[36] Woo, K., Lee, H.J., Ahn, J -P., Park, Y.S., 2003. Sol–Gel Mediated Synthesis of Fe2O3 Nanorods. Adv. Mater, 15, 1761–1764.

[37] Wang, Y., Teng, X., Wang, J-S., Yang, H., 2003. Solvent-Free Atom Transfer Radical Polymerization in the Synthesis of Fe2O3@Polystyrene Core−Shell Nanoparticles. Nano Lett., 3, 789–793.

[38] Hyeon, T., Lee, S.S., Park, J., Chung, Y., Na, H.B., 2001. Synthesis of Highly Crystalline and Monodisperse Maghemite Nanocrystallites without a Size-Selection Process. J. Am. Chem. Soc., 123, 12798–12801.

[39] Lu, Y., He, B., Shen, J., Li, J., Yang, W., Yin, M., 2015. Multifunctional magnetic and fluorescent core-shell nanoparticles for bioimaging. Nanoscale, 7, 1606–1609.

[40] Bao, F., Yao, J.L., Gu, R.A., 2009. Synthesis of Magnetic Fe2O3/Au core/shell nanoparticles for bioseparation and immunoassay based on surface-enhanced Raman spectroscopy. Langmuir, 25, 10782–10787.

[41] Caruso, F., Spasova, M., Susha, A., Giersig, M., Caruso, R.A., 2001. Magnetic Nanocomposite Particles and Hollow Spheres Constructed by a Sequential Layering Approach. Chem. Mater, 13, 109–116.

[42] Pankhurst, Q.A., Connolly, J., Jones, S.K., Dobson, J., 2003. Applications of magnetic nanoparticles in biomedicine. J. Phys. D. Appl. Phys., 36, R167.

[43] Meddahi-Pellé, A., Legrand, A., Marcellan, A., Louedec, L., Letourneur, D., Leibler, L., 2014. Organ Repair, Hemostasis, and In Vivo Bonding of Medical Devices by Aqueous Solutions of Nanoparticles. Angew. Chem Int. Ed., 53, 6369–6373.

[44] Tietze, R., Zaloga, J., Unterweger, H., Lyer, S., Friedrich, R.P., Janko, C., Pöttler, M., Dürr, S., Alexiou, C., 2015. Magnetic nanoparticle-based drug delivery for cancer therapy. Biochem. Biophys. Res. Commun., 468, 463–470.

[45] Mody, V.V., Cox, A., Shah, S., Singh, A., Bevins, W., Parihar, H., 2014. Magnetic nanoparticle drug delivery systems for targeting tumor. Appl. Nanosci., 4, 385–392.

[46] Chen, J., Guo, Z., Tian, H., Chen, X., 2016. Production and clinical development of nanoparticles for gene delivery. Mol. Ther. - Methods Clin. Dev., 3, 16023.

[47] Boyer, C., Whittaker, M.R., Bulmus, V., Liu, J., Davis, T.P., 2010. The design and utility of polymer-stabilized iron-oxide nanoparticles for nanomedicine applications. NPG Asia Mater, 2, 23–30.

[48] Kim, D.K., Zhang, Y., Kehr, J., Klason, T., Bjelke, B., Muhammed, M., 2001. Characterization and MRI study of surfactant-coated superparamagnetic nanoparticles administered into the rat brain. J. Magn. Magn. Mater, 225, 256–261.

[49] McCarthy, J.R., Weissleder, R., 2008. Multifunctional magnetic nanoparticles for targeted imaging and therapy. Adv. Drug Deliv. Rev., 60, 1241–1251.

[50] Yang, H-W., Hua, M-Y., Liu, H-L., Huang, C-Y., Wei, K-C., 2012. Potential of magnetic nanoparticles for targeted drug delivery. Nanotechnol. Sci. Appl., 5, 73–86.

[51] Goswami, M.M., Dey, C., Bandyopadhyay, A., Sarkar, D., Ahir, M., 2016. Micelles driven magnetite (Fe3O4) hollow spheres and a study on AC magnetic properties for hyperthermia application. J. Magn. Magn. Mater, 417, 376–381.

[52] Dey, C., Baishya, K., Ghosh, A., Goswami, M.M, Ghosh, A., Mandal, K., 2017. Improvement of drug delivery by hyperthermia treatment using magnetic cubic cobalt ferrite nanoparticles. J. Magn. Magn. Mater, 427, 168–174.

[53] Cabuil, V., 2004. Magnetic Nanoparticles:Preparation and Properties. Dekker Encyclopedia of Nanoscience and Nanotechnology, Chapter 119, Roldan group puplications.

[54] Morcos, S.K., 2007. Nephrogenic systemic fibrosis following the administration of extracellular gadolinium based contrast agents: is the stability of the contrast agent molecule an important factor in the pathogenesis of this condition? Br J Radiol, 80(950), 73-76.

[55] Ersoy, H., Rybicki, F.J., 2007. Biochemical safety profiles of gadolinium-based extracellular contrast agents and nephrogenic systemic fibrosis. J Magn Reson Imaging, 26(5), 1190-1197.

[56] Thakral, C., Alhariri, J., Abraham, J.L., 2007. Long-term retention of gadolinium in tissues from nephrogenic systemic fibrosis patient after multiple gadoliniumenhanced MRI scans: case report and implications. Constrast Media Mol Imaging, 2(4), 199-205.

[57] Muldoon, L.L., Sandor, M., Pinkston, K.E., Neuwelt, E.A., 2005. Imaging, distribution, and toxicity of superparamagnetic iron oxide magnetic resonance nanoparticles in the rat brain and intracerebral tumor. Neurosurgery, 57(4), 785-796.

[58] Moghimi, S.M., Hunter, A.C., Murray, J.C., 2001. Long circulating and target-specific nanoparticles: theory to practice. Pharmacol Rev., 53(2), 283-318.

[59] Sosnovik, D.E., Nahrendorf, M., Weissleder, R., 2007. Molecular magnetic resonance imaging in cardiovascular medicine. Circulation, 115(15), 2076-2086.

[60] Thorek, D.L.J., Chen, A., Czupryna, J., Tsourkas, A., 2006. Superparamagnetic iron oxide nanoparticle probes formolecularimaging. Ann Biomed Eng., 34(1), 23-38.

[61] Lopez-Quintela, M.A., Tojo, C., Blanco, M.C., Rio, L.G., Leis, J.R., 2004. Microemulsion dynamics and reactions in microemulsions. Curr Opin Colloid Interface Sci., 9(3-4), 264-278.

[62] Ayyub, P., Multani, M., Barma, M., Palkar, V.R., Vijayaraghavan, R., 1988. Size-induced structural phasetransitions and hyperfine properties of microcrystalline Fe2O3. J Phys C: Solid State Phys., 21(11), 2229-2245.

[63] Perez, J.A.L., Quintela, M.A.L., Mira, J., Rivas, J., Charles, S.W., 1997. Advances in the preparation of magnetic nanoparticles by the microemulsion method. J phys Chem B, 101(41), 8045-8047.

[64] Lee, Y., Lee, J., Bae, C.J., Park, J.G., Noh, H.J., et al. 2005. Largescale synthesis of uniform and crystalline magnetite nanoparticles using reverse micelles as nanoreactors under reflux conditions. Adv Funct Mater, 15(3), 503-509.

[65] Sjogren, C.E., Johansson, C., Naevestad, A., Sontum, P.C., BrileySaebo, K., et al. 1997. Crystal size and properties of superparamagnetic iron oxide (SPIO) particles. Magn Reson Imaging, 15(1), 55-67.

[66] Grancharov, S.G., Zeng, H., Sun, S.H., Wang, S.X., O'Brien, S., et al. 2005. Biofunctionalization of monodisperse magnetic nanoparticles and their use as biomolecular labels in a magnetic tunnel junction based sensor. J Phys Chem B, 109(26), 13030-13035.

[67] Jun, Y.W., Huh, Y.M., Choi, J.S., Lee, J.H., Song, H.T., et al. 2005. Nanoscale size effect of magnetic nanocrystals and their utilization for cancer diagnosis via magnetic resonance imaging. J Am Chem Soc., 127(16):5732-5733.

[68] Khan, G.G., Sarkar, D., Singh, A.K., Mandal, K., 2013. Enhanced band gap emission and ferromagnetism of Au nanoparticle decorated [small alpha]-Fe2O3 nanowires due to surface plasmon and interfacial effects. RSC Adv., 3, 1722–1727.

[69] Cho, S-J., Jarrett, B.R., Louie, A.Y., Kauzlarich, S.M., 2006. Gold-coated iron nanoparticles: a novel magnetic resonance agent for T 1 and T 2 weighted imaging. Nanotechnology, 17, 640.

[70] Qu, L.H., Peng, Z.A., Peng, X.G., 2001. Alternative routes toward high quality CdSe nanocrystals. Nano Lett., 1(6), 333-337.

[71] Hines, M.A., Guyot-Sionnest, P., 1996. Synthesis and characterization of strongly luminescing ZnS-Capped CdSe nanocrystals. JPhysChem., 100(2), 468-471.

[72] Zhang, M., de Respinis, M., Frei, H., 2014. Time-resolved observations of water oxidation intermediates on a cobalt oxide nanoparticle catalyst. Nat. Chem., 6 (4), 362-367.

[73] Xiao, J., Tian, X.M., Yang, C., Liu, P., Luo, N.Q., Liang, Y., Li, H.B., Chen, D.H., Wang, C.X., Li, L., Yang, G.W., 2013. Ultrahigh relaxivity and safe probes of manganese oxide nanoparticles for in vivo imaging. Sci. Rep., 3, 3424.

[74] Li, C., Han, X., Cheng, F., Hu, Y., Chen, C., Chen, J., 2015. Phase and composition controllable synthesis of cobalt manganese spinel nanoparticles towards efficient oxygen electrocatalysis. Nat. Commun., 6, 7345.

[75] Zhao, Z., Zhou, Z., Bao, J., Wang, Z., Hu, J., Chi, X., Ni, K., Wang, R., Chen, X., Chen, Z., Gao, J., 2013. Octapod iron oxide nanoparticles as high-performance T2 contrast agents for magnetic resonance imaging. Nat. Commun., 4, 2266.

[76] Bethe, H.A., 1929. Splitting of Terms in Crystals. Ann. Phys., 3, 133–206.

[77] Néel, L., 1954. Anisotropie magnétique superficielle et surstructures d'orientation. J. Phys. Radium, 15, 225–239.

[78] Skomski, R., Coey J.M.D., 1999. Permanent magnetism. (Bristol, UK ; Philadelphia, PA : Institute of Physics Pub.

[79] N. I. for Materials Science, "Magnetoelastic anisotropy," visited 24-March-2015. [Online]. Available: http://www.nims.go.jp/apfim/pdf/ MMC_Lecture4.pdf.

[80] Bland, J., 2003. visited 24-March-2015. [Online]. Available: http: //www.cmp.liv.ac.uk/frink/thesis/thesis/node70.html.

[81] Cullity, B.D., Graham, C.D., 2008. Introduction to Magnetic Materials 2nd Edition. Wiley–IEEE, ISBN 978-0-471-47741-9.

[82] Cullity, B.D., 1972. Introduction to magnetic materials. Reading, Mass., Addison-Wesley Pub. Co.

[83] Wu, W., He, Q., Jiang, C., 2008. Magnetic iron oxide nanoparticles: Synthesis and surface functionalization strategies. Nanoscale Res. Lett., 3, 397–451.

[84] Zhu, N., Ji, H., Yu, P., Niu, J., Farooq, M.U., Akram, M.W., Udego, I. O., Li, H., Niu, X., 2018. Surface modification of magnetic iron oxide nanoparticles. Nanomaterials, 8, 810.

[85] Pušnik, K., Goršak, T., Drofenik, M., Makovec, D., 2016. Synthesis of aqueous suspensions of magnetic nanoparticles with the coprecipitation of iron ions in the presence of aspartic acid. J. Magn. Magn. Mater., 413, 65-75.

[86] Hui, C., Shen, C., Tian, J., Bao, L., Ding, H., Tian, Y., Shi, X., Gao, H., 2011. Core–shell Fe3O4@SiO2 nanoparticles synthesized with well-dispersed hydrophilic Fe3O4 seeds. Nanoscale, 3, 701-705.

[87] Sonmez, M., Georgescu, M., Alexandrescu, L., Gurau, D., Ficai, A., Ficai, D., Andronescu, E., 2015. Synthesis and applications of Fe3O4/SiO2 core–shell materials. Curr. Pharm. Des., 21, 5324-5335.

[88] Li, F., Yu, Z., Zhao, L., Xue, T., 2016. Synthesis and application of homogeneous Fe3O4 core/Au shell nanoparticles with strong SERS effect. RSC Adv., 6, 10352-10357.

[89] Lee, M.H., Leu, C.C., Lin, C.C., Tseng, Y.F., Lin, H.Y., Yang, C. N., 2019. Gold-decorated magnetic nanoparticles modified with hairpin-shaped DNA for fluorometric discrimination of singlebase mismatch DNA. Microchim. Acta, 186 (2), 80.

[90] Zhao, J., Tu, K., Liu, Y., Qin, Y., Wang, X., Qi, L., Shi, D., 2017. Photo-controlled aptamers delivery by dual surface goldmagnetic nanoparticles for targeted cancer therapy. Mater. Sci. Eng.C, 80, 88-92.

[91] Chen, S., Liu, N., Yanyun, J., Xiong, C., Dong, L., 2017. In situ synthesis and antibacterial application of Fe3O4@Ag nanoparticles. J. Func. Mater., 48, 03097.

[92] Portet, D., Denizot, B., Rump, E., Lejeune, J., Jallet, P., 2001. Nonpolymeric coatings of iron oxide colloids for biological use as magnetic resonance imaging contrast agents. J. Colloid Interface Sci., 238 (1), 37-42.

[93] Soares, P., Laia, C., Carvalho, A., Pereira, L., Coutinho, J., Ferreira, I., Novo, C., Borges, J., 2016. Iron oxide nanoparticles stabilized with a bilayer of oleic acid for magnetic hyperthermia and MRI applications. Appl. Surf. Sci., 383, 240-247.

[94] Andreas, K., Georgieva, R., Ladwig, M., Mueller, S., Notter, M., Sittinger, M., Ringe, J., 2012. Highly efficient magnetic stem cell labeling with citrate-coated superparamagnetic iron oxide nanoparticles for MRI tracking. Biomaterials, 33 (18), 4515-4525.

[95] Ardelean, I., Stoencea, L., Ficai, D., Ficai, A., Trusca, R., Vasile, B., Nechifor, G., Andronescu, E., 2017. Development of stabilized magnetite nanoparticles for medical applications. J. Nanomater. 2017, ID 6514659.

[96] Yamaura, M., Camilo, R., Sampaio, L., Macedo, M., Nakamura, M., Toma, H., 2004. Preparation and characterization of (3-aminopropyl) triethoxysilane-coated magnetite nanoparticles. J. Magn. Magn. Mater. 279, 210-217.

[97] Sun, X., Li, Y., 2013. Functional modification and preparation of superparamagnetic Fe3O4. Adv. Mater. Res., 743, 183-188.

[98] Xiong, Z., Li, S., Xia, Y., 2016. Highly stable water-soluble magnetic nanoparticles synthesized through combined co-precipitation, surface-modification, and decomposition of a hybrid hydrogel. New J. Chem., 40, 9951-9957.

[99] Shahidi, F., Abuzaytoun, R., 2005. Chitin, chitosan, and coproducts: Chemistry, production, applications, and health effects. Adv. Food Nutr. Res., 49, 93-135.

[100] Radwan, M.A., Rashad, M.A., Sadek, M.A., Elazab, H.A., 2019. Synthesis, characterization and selected application of chitosan coated magnetic iron oxide nanoparticles. J. Chem. Technol. Metall., 54 (2), 303-310.

[101] Boustani, K., Shayesteh, S., Salouti, M., Jafari, A., Shal, A., 2018. Synthesis, characterisation and potential biomedical applications of magnetic core–shell structures: Carbon-, dextran-, SiO2- and ZnO-coated Fe3O4 nanoparticles. New J. Chem., 12 (1), 78-86.

[102] Zhao, X., Cui, H., Chen, W., Wang, Y., Cui, B., Sun, C., Meng, Z., Liu, G., 2014. Morphology, structure and function characterization of PEI modified magnetic nanoparticles gene delivery system. PLoS One, 9 (6), e98919.

[103] Patsula, V., Tulinska, J., Trachtová, Š., Kuricova, M., Liskova, A., Španová, A., Ciampor, F., Vavra, I., Rittich, B., Ursinyova, M., Dusinska, M., Ilavska, S., Horvathova, M., Masanova, V., Uhnakova, I., Horák, D., 2019. Toxicity evaluation of monodisperse PEGylated magnetic nanoparticles for nanomedicine. Nanotoxicology, 13 (4), 1-17.

[104] Park, H., Mcconnell, J., Boddohi, S., Kipper, M., Johnson, P., 2011. Synthesis and characterization of enzyme-magnetic nanoparticle complexes: Effect of size on activity and recovery. Colloids Surf., B 83 (2), 198-203.

[105] Riahi, R., Tamayol, A., Shaegh, S., Ghaemmaghami, A., Dokmeci, M., Khademhosseini, A., 2015. Microfluidics for advanced drug delivery systems. Curr. Opin. Chem. Eng., 7, 101-112.

[106] Iravani, S.,2011. Green synthesis of metal nanoparticles using plants. Green Chem.,13, 2638-2650.

[107] Gul, S., Khan, S.B., Rehman, I.U., Khan, M.A., Khan, M.I., 2019. A Comprehensive Review of Magnetic Nanomaterials Modern Day Theranostics. Front. Mater., 6, 179.

[108] Choi, H.S., Liu, W., Liu, F., Nasr, K., Misra, P., Bawendi, M.G., Frangioni, J.V., 2010. Design considerations for tumour-targeted nanoparticles. Nat. Nanotechnol., 5 (1), 42–47.

[109] Laurent, S., Forge, D., Port, M., Roch, A., Robic, C., Elst, L.V., Muller, R.N., 2008. Magnetic iron oxide nanoparticles: Synthesis, stabilization, vectorization, physicochemical characterizations, and biological applications. Chem. Rev., 108(6), 2064-2110.

[110] Park, J.H., Maltzahn, G.V., Zhang, L., Derfus, A.M., Simberg, D., Harris, T.J., Ruoslahti, E., Bhatia, S.N., Sailor, M.J., 2009. Systematic surface engineering of magnetic nanoworms for in vivo tumor targeting. Small, 5 (6), 694–700.

[111] Kohler, N., Fryxell, G.E., Zhang, M., 2004. A Bifunctional Poly(ethylene glycol) Silane Immobilized on Metallic Oxide-Based Nanoparticles for Conjugation with Cell Targeting Agents. J. Am. Chem. Soc., 126 (23), 7206–7211.

[112] Shen, T., Weissleder, R., Papisov, M., Bogdanov, A., Brady, T.J., 1993. Monocrystalline iron oxide nanocompounds (MION): physicochemical properties. Magn. Reson. Med., 29 (5), 599–604.

[113] Ho, K.M., Li, P., 2008. Design and Synthesis of Novel Magnetic Core−Shell Polymeric Particles. Langmuir, 24 (5), 1801–1807.

[114] Plassat, V., Martina, M.S., Barratt, G., Menager, C., Lesieur, S., 2007. Sterically stabilized superparamagnetic liposomes for MR imaging and cancer therapy: Pharmacokinetics and biodistribution. Int. J. Pharm., 344, 118–127.

[115] McBain, S.C., Yiu, H.H.P., Haj, A.E., Dobson, J., 2007. Polyethyleneimine functionalized iron oxide nanoparticles as agents for DNA delivery and transfection. J. Mater. Chem., 17, 2561–2565.

[116] Hao, R., Xing, R., Xu, Z., Hou, Y., Gao, S., Sun, S., 2010. Synthesis, Functionalization, and Biomedical Applications of Multifunctional Magnetic Nanoparticles. Adv. Mater., 22, 2729–2742.

[117] Mornet, S., Vasseur, S., Grasset F., Duguet, E., 2004. Magnetic nanoparticle design for medical diagnosis and therapy. J. Mater.Chem., 2004, 14, 2161–2175.

[118] Kader, R.A., Rose, L.C., Suhaimi, H., Manickam, M.S., 1885. Synthesis and toxicity test of magnetic nanoparticle via biocompatible microemulsion system as template for application in targeted drug delivery. AIP Conf. Proc., 1885, 020136.

[119] Fan, C., Gao, W., Chen, Z., Fan, H., Li, M., 2010. Tumor selectivity of stealth multi-functionalized superparamagnetic iron oxide nanoparticles. Int. J. Pharm., 404 (1-2), 180-190.

[120] Jang, B., Park, S., Kang, S., Kim, J., Kim, S., 2012. Gold nanorods for target selective SPECT/CT imaging and photothermal therapy in vivo. Quant. Imag. Med. Surg., 2 (1), 1-11.

[121] Shellman, Y.G., Howe, W.R., Miller, L.A., Goldstein, N.B., Pacheco, T.R., Mahajan, R.L., LaRue, S.M., Norris, D.A., 2008. Hyperthermia induces endoplasmic reticulum-mediated apoptosis in melanoma and non-melanoma skin cancer cells. J. Invest. Dermatol., 128, 949–956.

[122] Hildebrandt, B., Wust, P., Ahlers, O., Dieing, A., Sreenivasa, G., Kerner, T., Felix R., Riess, H., 2002. The cellular and molecular basis of hyperthermia.Crit. Rev. Oncol. Hemat., 43 (1), 33–56.

[123] Thomas, L., Dekker, L., Kallumadil, M., 2009. Carboxylic acid-stabilised iron oxide nanoparticles for use in magnetic hyperthermia. J. Mater. Chem., 19, 6529-6535.

[124] Moroz, P., Jones, S., Gray, B., 2001. Status of hyperthermia in the treatment of advanced liver cancer. J. Surg. Oncol., 77 (4), 259-269.

[125] Kubes, J., Svoboda, J., Rosina, J., Starec M., Fiserova, A., 2008. Immunological response in the mouse melanoma model after local hyperthermia.Physiol. Res., 2008, 57 (3), 459–465.

[126] Wust, P., Hildebrandt, B., Sreenivasa,G., Rau, B., Gellermann, J., Riess, H., Felix R., Schlag, P. M., 2002. Hyperthermia in combined treatment of cancer. Lancet Oncol., 3 (8), 487–497.

[127] Falk, M.H., Issels, R.D., 2001. Hyperthermia in oncology. Int. J. Hyperthermia, 17 (1), 1–18.

[128] Kapp, D.S., Hahn, G.M., Carlson, R.W., 2000. Principles of Hyperthermia. 5th edn, B.C. Decker Inc., 2000.

[129] McMahon, H.T., Gallop, J.L., 2005. Membrane curvature and mechanisms of dynamic cell membrane remodelling. Nature, 438(7068), 590-596.

[130] Shi, Y., Massagué, J., 2003. Mechanisms of TGF-β signaling from cell membrane to the nucleus. Cell, 113(6), 685–700.

[131] García-Sánchez, T., Muscat, A., Leray, I., Mir, L.M., 2018. Pyroelectricity as a possible mechanism for cell membrane permeabilization. Bioelectrochemistry, 119, 227–233.

[132] Honigmann, A., Pralle, A., 2016. Compartmentalization of the Cell Membrane. J Mol Biol., 428(24, Part A), 4739–4748.

[133] Zalba, S., Hagen, T.L.M.T., 2017. Cell membrane modulation as adjuvant in cancer therapy. Cancer Treat Rev., 52, 48–57.

[134] Panariti, A., Miserocchi, G., Rivolta, I., 2012. The effect of nanoparticle uptake on cellular behavior: disrupting or enabling functions? Nanotechnol. Sci. Appl., 5, 87–100.

[135] Clift, M.J.D., Brandenberger, C., Rothen-Rutishauser, B., Brown, D.M., Stone, V., 2011. The uptake and intracellular fate of a series of different surface coated quantum dots in vitro. Toxicology, 286 (1-3), 58–68.

[136] Chen, J., Yu, Z., Chen, H., Gao, J., Liang, W., 2011. Transfection efficiency and intracellular fate of polycation liposomes combined with protamine. Biomaterials, 32 (5), 1412–1418.

[137] Martínez-Riaño, A., Bovolenta, E.R., Mendoza, P., Oeste, C.L., Martín-Bermejo, M.J., Bovolenta, P., Turner, M., Martínez-Martín, N., Alarcón, B., 2018. Antigen phagocytosis by B cells is required for a potent humoral response. EMBO Reports, 19 (9), e46016.

[138] Chen, F., Wang, G., Grif, J.I., Brenneman, B., Banda, N.K., Holers, V.M., Backos, D.S., Wu, L., Moghimi, S.M., Simberg, D., 2017. Complement proteins bind to nanoparticle protein corona and undergo dynamic exchange in vivo. Nat. Nanotechnology, 12, 387-393.

[139] Tavano, R., Gabrielli, L., Lubian, E., Fedeli, C., Visentin, S., Laureto, P.P.D., Arrigoni, G., Ge, A., Chen, F., Simberg, D., Morgese, G., Benetti, E.M., Wu, L., Moghimi, S.M., Mancin, F., Papini, E., 2018. C1q-mediated complement activation and C3 opsonization trigger recognition of stealth poly(2-methyl-2-oxazoline)-coated silica nanoparticles by human phagocytes. ACS Nano, 12 (6) 5834–5847.

[140] Walkey, C.D., Olsen, J.B., Guo, H., Emili, A., Chan, W.C.W., 2012. Nanoparticle size and surface chemistry determine serum protein adsorption and macrophage uptake. J. Am. Chem. Soc., 134 (4), 2139–2147.

[141] Dai, Q., Walkey, C., Chan, W.C.W., 2014. Polyethylene glycol backfilling mitigates the negative impact of the protein corona on nanoparticle cell targeting. Angew. Chem. Int. Ed., 53 (20), 5093–5096.

[142] Li, Y., Kröger, M., Liu, W.K., 2014. Endocytosis of PEGylated nanoparticles accompanied by structural and free energy changes of the grafted polyethylene glycol. Biomaterials, 35 (30), 8467–8478.

[143] Saha, K., Kim, S.T., Yan, B., Miranda, O.R., Alfonso, F.S., Shlosman, D., Rotello, V.M., 2013. Surface Functionality of Nanoparticles Determines Cellular Uptake Mechanisms in Mammalian Cells. Small, 9 (2), 300–305.

[144] Chithrani, B.D., Ghazani, A.A., Chan, W.C.W., 2006. Determining the size and shape dependence of gold nanoparticle uptake into mammalian cells. Nano Lett. 6, 662–668.

[145] Yang, K., Ma, Y.Q., 2010. Computer simulation of the translocation of nanoparticles with different shapes across a lipid bilayer. Nat. Nanotechnol., 5, 579–583.

[146] Dasgupta, S., Auth, T., Gompper, G., 2014. Shape and orientation matter for the cellular uptake of nonspherical particles. Nano Lett., 14, 687–693.

[147] Jiang, Y., Huo, S., Mizuhara, T., Das, R., Lee, Y.-W., Hou, S., Moyano, D. F., Duncan, B., Liang, X.-J., Rotello, V. M., 2015. The Interplay of Size and Surface Functionality on the Cellular Uptake of Sub-10 nm Gold Nanoparticles. ACS Nano, 9, 9986−9993.

[148] Teja, A.S., Koh, P-Y., 2009. Synthesis, properties, and applications of magnetic iron oxide nanoparticles. Progress in Crystal Growth and Characterization of Materials, 55 (1–2), 22–45.

[149] Bulte, J.W.M., Kraitchman, D.L., 2004. Iron oxide MR contrast agents for molecular and cellular imaging. NMR in Biomedicine, 17 (7), 484–499.

[150] Javidi, M., Heydari, M., Attar, M.M., Haghpanahi, M., Karimi, A., Navidbakhsh, M., Amanpour, S., 2015. Cylindrical agar gel with fluid flow subjected to an alternating magnetic field during hyperthermia. International Journal of Hyperthermia, 31 (1), 33–39.

[151] Javidi, M., Heydari, M., Karimi, A., Haghpanahi, M., Navidbakhsh, M., Razmkon, A., 2014. Evaluation of the effects of injection velocity and different gel concentrations on nanoparticles in hyperthermia therapy. J Biomed Phys Eng., 4 (4), 151–162.

[152] Heydari, M., Javidi, M., Attar, M.M., Karimi, A., Navidbakhsh, M., Haghpanahi, M., Amanpour, S., 2015. Magnetic Fluid Hyperthermia in a Cylindrical Gel Contains Water Flow. Journal of Mechanics in Medicine and Biology, 15 (5), 1550088.

[153] Manju, K., Smitha, T., Divya, S.N., Aswathy, E.K., Aswathy, B., Arathy, T., Binu-Krishna, K.T., 2015. Structural, magnetic, and acidic properties of cobalt ferrite nanoparticles synthesised by wet chemical methods. J. Adv. Ceram., 4, 199–205.

[154] Swatsitang, E., Phokha, S., Hunpratub, S., Usher, B., Bootchanont, A., Maensiri, S., Chindaprasirt, P., 2016. Characterization and magnetic properties of cobalt ferrite nanoparticles. J. Alloys Compd., 664, 792–797.

[155] Sharifi, I., Shokrollahi, H., Amiri, S., 2012. Ferrite-based magnetic nanofluids used in hyperthermia applications. J. Magn. Magn. Mater, 324(6), 903–915.

[156] Amiri, S., Shokrollahi, H., 2013. The role of cobalt ferrite magnetic nanoparticles in medical science. Mater. Sci. Eng. C, 33(1), 1–8.

[157] Mathew, D.S., Juang, R.S., 2007. An overview of the structure and magnetism of spinel ferrite nanoparticles and their synthesis in microemulsions. Chemical Engineering Journal, 129 (1-3), 51-65.

[158] Bueno, A.R., Gregori, M.L., Nobrega, M.C.S., 2007. Ultrafast dynamics of 1 μm ZnO epitaxial films by time-resolved measurements. Mater. Chem. Phys., 105, 229–233.

[159] Sagadevan, S., Podder, J., Das, I., 2017. Synthesis and characterization of cobalt ferrite (CoFe2O4) nanoparticles prepared by hydrothermal method. Recent Trends Mater. Sci. Appl., 145–152.

[160] Ding, Z., Wang, W., Zhang, Y., Li, F., Ping-Liu, J., 2015. Synthesis, characterization and adsorption capability for Congo red of CoFe2O4 ferrite nanoparticles. J. Alloy. Compd., 640, 362–370.

[161]. Kefeni, K.K., Mamba, B.B., Msagati, T.A.M., 2017. Application of spinel ferrite nanoparticles in water and wastewater treatment: a review. Sep. Purif. Technol., 188, 399–422.

[162]. Liu, F., Laurent, S., Roch, A., Vander Elst, L., Muller, R.N., 2013. Size-controlled synthesis of CoFe2O4 nanoparticles potential contrast agent for MRI and investigation on their size-dependent magnetic properties. J. Nanomater. 2013 , 1–9.

[163]. Lee, S.W., Bae, S., Takemura, Y., et al. 2007. Self-heating characteristics of cobalt ferrite nanoparticles for hyperthermia application. J. Magn. Magn. Mater, 310, 2868–2870.

[164]. Fantechi, E., Innocenti, C., Albino, M., Lottini, E., Sangregorio, C., 2015. Influence of cobalt doping on the hyperthermic efficiency of magnetite nanoparticles. J. Magn. Magn. Mater, 380, 365–371.

[165] Mohaideen, K.K., Joy, P.A., 2012. High magnetostriction and coupling coefficient for sintered cobalt ferrite derived from superparamagnetic nanoparticles. Appl. Phys. Lett., 101 72405.

[166] Liu, C., Zou, B., Rondinone, A.J., Zhang, Z.J., 2000. Chemical Control of Superparamagnetic Properties of Magnesium and Cobalt Spinel Ferrite Nanoparticles through Atomic Level Magnetic Couplings. J. Am. Chem. Soc., 122 6263–6267.

[167] Kline, T.L., Xu, Y-H., Jing, Y., Wang, J-P., 2009. Biocompatible high-moment FeCo-Au magnetic nanoparticles for magnetic hyperthermia treatment optimization. Journal of Magnetism and Magnetic Materials, 321, 1525–1528.

[168] Song, Q., Zhang, Z.J., 2004. Shape Control and Associated Magnetic Properties of Spinel Cobalt Ferrite Nanocrystals. J. Am. Chem. Soc., 126, 6164–6168.

[169] Pal, D., Mandal, M., Chaudhuri, A., Das, B., Sarkar, D., Mandal, K., 2010. Micelles induced high coercivity in single domain cobalt-ferrite nanoparticles. J. Appl. Phys., 108.

Chapter 2

Experimental details

In this chapter we have discussed the synthesis procedure of different types of stable magnetic nanoparticles (MNPs) which usually we have used in different purposes throughout our research investigation to explore their applicability in many fields. Here, we also have discussed about different types of characterization techniques such as X-Ray Diffraction (XRD), Transmission Electron Microscope (TEM), Scanning electron microscope (SEM), High Resolution Transmission Electron Microscope (HRTEM), Ultraviolet-visible (UV-Vis) spectroscopy, Fourier Transformed Infrared (FTIR), Vibrating Sample Magnetometer (VSM), Superconducting Quantum Interference Device (SQUID), Isothermal Titration Calorimetry (ITC) by which we can investigate the physical, structural, morphological and magnetic properties of the prepared nanomaterials. We also have discussed about the short introduction on the different types of instruments which we have used in our research purpose.

2.1. Induction:

In this chapter I have focused on different experimental procedures such as wet chemical and chemical co-precipitation, solvothermal methods which have used for synthesis of transition metal based magnetic nanomaterials (NMs) i.e. magnetite (Fe_3O_4), cobalt ferrite ($CoFe_2O_4$) and manganese ferrite ($MnFe_2O_4$) nanoparticle (NP). This chapter also gives a brief overview of characterization processes of these NPs to observe their chemical, physical, optical and magnetic properties.

The morphology, size and microstructure of these NMs were studied using XRD, TEM, SEM and HRTEM. Surface chemistry of NMs was studied by spectroscopic data of UV-Vis spectroscopy, FTIR and fluorescence microscope. The magnetic properties of NMs were checked by VSM and SQUID whereas; AC magnetic measurements are performed by a lab made setup by which we measured the high frequency magnetic behavior under an AC magnetic field. The binding affinity between two molecules was observed by ITC. We have prepared a basic solenoid with 174 no of turns operating at 230 V with 50 Hz AC supply where we measured the heating property of NMs in hyperthermic condition. Here, we also discuss about some other instruments which were used for our work like, centrifuge, sonicator, magnetic stirrer etc.

2.2. Synthesis of nanomaterials:

Generally two approaches are used for synthesis of metallic nanoparticles:

a) Top-down approach
b) Bottom-up approach.

2.2.1. Top-Down approach

In this approach the bulk materials are breaking down into nanosized structures and nanoparticles form with wide size range and various shapes. Nanoparticles are synthesized by this method by chemical etching, laser or thermal ablation, sonication sputtering etc. The biggest drawback of the top-down approach is the imperfection of surface structure of nanomaterial. Nanowires made by lithography is one of the example of top-down process may contain many impurities and structural defects. This technique is used in high energy wet ball milling, atomic force manipulation, gas-phase condensation, electron beam lithography, computer hard disk etc.

2.2.2. Bottom-up approach

In bottom-up approach, the nanomaterial is build from bottom level i.e.; atom by atom, molecule by molecule or cluster by cluster. By this method well defined nanostructures are formed with fewer defects by spontaneous organization of the atoms and molecules. Some examples of bottom-up technique are sol-gel synthesis, colloidal precipitation, oregano-metallic chemical route, revere-micelle route, hydrothermal synthesis, electro-deposition etc which are used for synthesis of luminescent nanoparticles. Bottom-up approach is easier, cost friendly and more controlled than top- down method. Schematic diagram of top-down and bottom-up approach are shown in Fig.2.1.

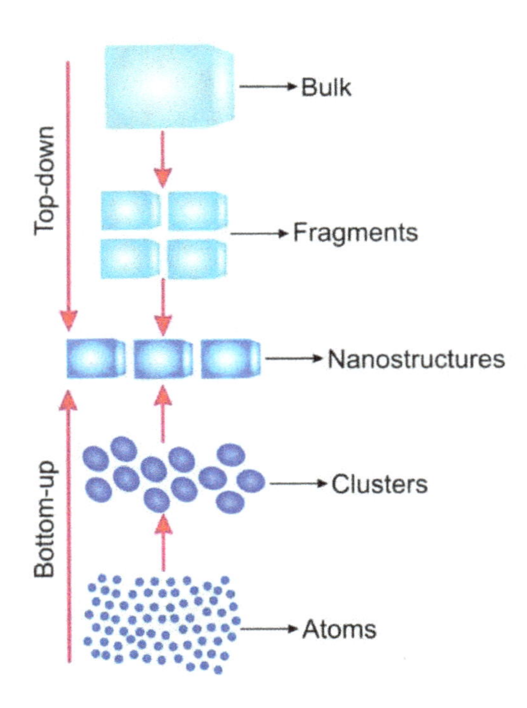

Source: https://images.app.goo.gl/AUMviK5GjAMhdTHV6

Fig.2.1. Schematic representation for synthesis of nanoparticles.

2.3. Fabrication techniques of various nanoparticles:

2.3.1. Wet-chemical method:

In wet-chemical method, the chemical reactions occur in the solution phase using precursors. This method is mainly a bottom-up approach. By this method, 2D nanomaterials have been synthesized that are unable to prepare by top-down approach. The main wet-chemical synthesis routes for 2D nanomaterials are solvothermal synthesis, self-assembly, template synthesis, hot-injection, oriented attachment and interface mediated synthesis.

2.3.2. Chemical co-precipitation:

In co-precipitation process two or more substances are formed simultaneously. The substances are usually soluble in solvent and under chemical procedure they generate a single compound. This method is generally used to synthesize of iron oxide nanoparticles in material science with high crystallinity and better stoichiometry. For synthesis of iron oxide nanoparticles by this technique, iron based salts are take with proper molar ratio and are dissolved in aqueous solution and the solution stirred magnetically and heated for proper mixing. As soon as the temperature of the solution reached around 85°C, a basic solution like sodium hydroxide (NaOH) or potassium hydroxide (KOH) is added drop by drop into the following chemical reaction which helps to maintain the solution pH. The solution's temperature is sustained at this same temperature for 1 h. The size, shape and composition of the synthesized particles are dependent on temperature, pH, ionic strength, kind of basic solution etc of the solution. Fig.2.2. shows a schematic diagram of co-precipitation method for nanomaterial synthesis. We have synthesized the magnetite and cobalt ferrite nanoparticles by this co-precipitation method.

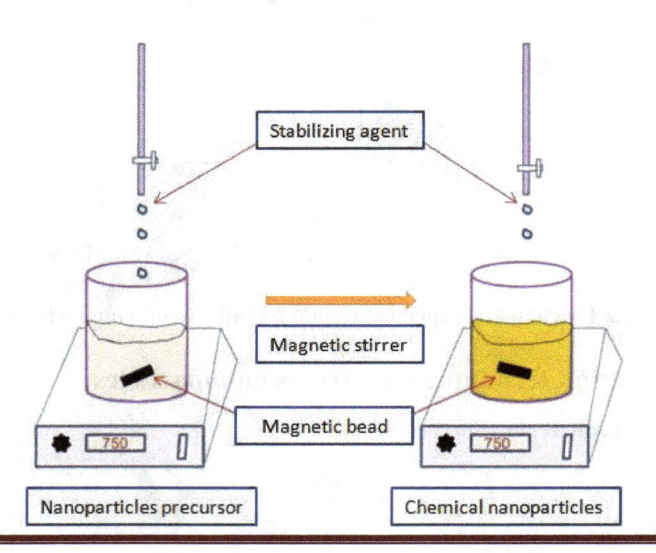

https://images.app.goo.gl/opfkSnAFSMr8auKx6

Fig.2.2. Schematic diagram of co-precipitation method for synthesis of nanomaterials.

2.3.3. Solvothermal method:

It is a method by which different nanomaterials i.e.; various metals, semiconductors, polymers, ceramics can be synthesized in an autoclave by using different solvents such as polar (water, ethanol etc.) and non-polar solvents (benzene, ethylene glycol etc.) under high pressure and temperature. In a typical solvothermal process, the solvents act as the reaction medium in a closed vessel. Here, the reaction temperature should be higher than the boiling point of solvent and the solvent will be self generated at high pressure that improves the crystallinity of the synthesized nanomaterials [1]. Two-dimensional nanosheets of metals [2], metal oxides [3], and metal chalcogenides [4,5] have been reported which are synthesized by this method. In solvothermal method, the nanoparticles are obtained by the dissolution and crystallization mechanism. At higher temperature, the solubility and activity of the reactants changes that give some additional parameters whose controlled tuning causes formation of nanoparticles with high quality. Fig.2.3. shows the schematic representation of synthesis of nanoparticles by solvothermal method. This technique starts at room temperature when the precursor's molecules are dissolved in the solvents by magnetic stirring. Then the reaction mixture is poured into a Teflon lined stainless steel chamber in such a way that filled 80% of the Teflon chamber and put it into an autoclave. The autoclave is then heated to a certain temperature which creates two different temperature regions inside the oven. Therefore, the solvents settled at below gets heated firstly which shifted up and the solvent from the top which remain cold passed down. At bottom part of the solution, the nutrients get dissolved due to heat and the saturated solution of this lower part is then transferred to the upper portion of the solution and this process goes until the whole solution reaches at its equilibrium. When the solution gets supersaturated in the upper part, then the crystallization of the sample starts. By this method nanomaterials have synthesized with different shapes like rods; hollow spheres etc. are shown in Fig.2.4. Different thermodynamically stable novel materials can also be synthesized by this technique which is not easy to synthesize by other techniques. More logics are present for formation of these particles such as selective adsorption of surfactants [6], oriented attachment of phases [7,8,9], minimization of surface free energy [10] and molecular template mechanism [7].

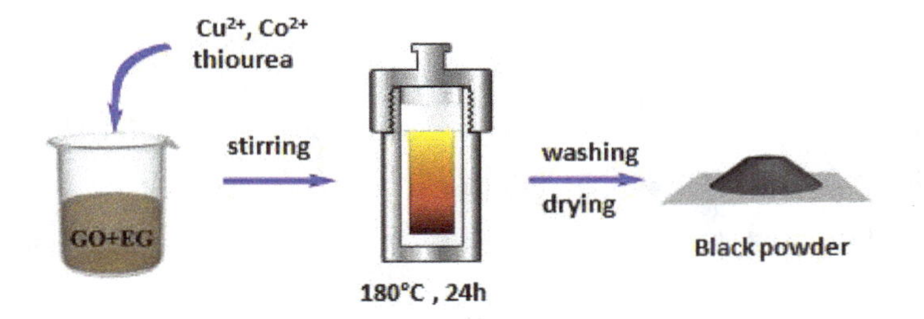

https://images.app.goo.gl/FcKQc2uwWQZMHMRcA

Fig.2.3. Schematic diagram of synthesis of nanoparticles by solvothermal method.

Fig.2.4. SEM image of rod and hollow shaped nanoparticles synthesized by solvothermal method.

2.4. Morphological characterization techniques:

2.4.1. X-ray crystallography:

X-ray crystallography is a method of determining the arrangement of atoms with in a crystal, in which a beam of X- rays hit the crystallize molecule and causes the beam of light to diffract into multiple specific directions depending on the crystal's atomic orientation. The angles and intensities of these diffracted beams is analogue to spatial arrangement of atom in crystal which produce a three dimensional picture of the density of electrons present within the crystal. Fig.2.5.(a) shows a schematic representation of a typical X-ray diffractometer. A typical diffractometer consist of following parts- X-ray source, monochromater, calorimeter, goniometer, photographic plate film and detector. When high energy electron beam is strikes on the sample, the inner shell electrons tears out and the electrons from the next excited states fill up the vacancy which causes generation of X-ray. From the X-ray source the monochromatic X-ray beam fall on a crystalline sample which then scattered elastically by the electrons within the crystal planes and intervene constructively in few specific directions determined by Bragg's law:

$$2d \sin\Theta = n\lambda$$

Where n is a positive integer, λ is the wavelength of the incident wave, d is the spacing between two consecutive parallel crystallographic planes aligned in the same direction and Θ is the angle of diffraction . Detector records the angles and intensities of the diffracted beams. In X- ray diffraction pattern, the intensity is plotted against the angle 2Θ. The X-ray diffraction pattern of cobalt ferrite nanoparticle is shown in Fig.2.5.(b). The size (D) of the particle can be calculating from the width of the diffraction peaks by Debye-Scherrer formula as given below:

$$D = \frac{K\lambda}{\beta cos\theta}$$

Where, K is a constant depends on the shape of the particle; β is the full width at the half maxima of the diffraction peak at angle 2Θ.

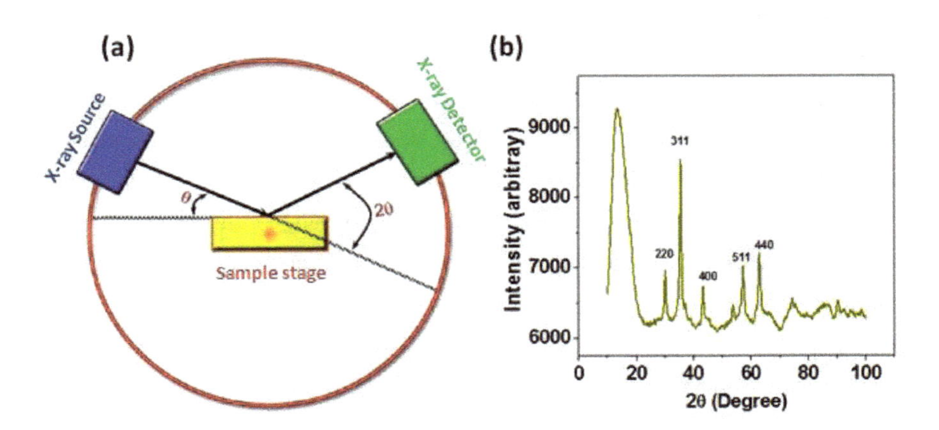

https://images.app.goo.gl/mCF4LArVPiYb6kn68

Fig.2.5.(a) Schematic diagram of XRD in Θ-2Θ mode, (b) X-ray diffraction pattern of cobalt ferrite nanoparticle.

2.4.2. Electron Microscopes

An electron microscope is a microscope that uses a high energetic electron beam which illuminates a specimen and produced its magnified image. In our work two types of electron microscope have been used are discussed below.

2.4.2.1. Scanning Electron Microscopy:

A scanning electron microscope (SEM) is a type of electron microscope where the sample image was produced by scanning the surface of sample with a focused beam of electrons. The electrons in the beam interact with the sample and produce various signals which give the information about the sample's composition, surface topography and other properties such as electrical conductivity. The SEM has a large depth of field, which allows a large amount of the sample to be focused at one time and produces an image that is a good representation of the three- dimensional sample. The beam of electrons are emitted from a cathode filament made of a thin tungsten wire (~0.1mm) by heating the filament at high temperature (~2800k) and passes through condenser lenses, scan coils and objective lens and finally enters into the sample chamber. In SEM, the "Virtual Source" at the top represents the electron gun which producing a stream of monochromatic electrons. The stream is then condensed by the first condenser lens which limits the amount of current in the beam. The beam is then constricted by condenser aperture by eliminating some high angle electrons from the beam and enters into the second condenser lens. The second condenser lens forms the electrons into a thin, tight, coherent beam and passes through objective aperture which further eliminates the high angle electrons from the beam. Then a set of coils scan the beam in a grid fashion and the beam is then enters into the objective lens which focuses the scanning beam onto the sample. This electron-sample interaction produced a variety of signals which include secondary electrons, backscattered electrons, diffracted backscattered electrons, characteristic X-rays, visible light and heat (Fig.2.7.). Secondary electrons and backscattered electrons are commonly used for imaging samples. Secondary electrons are most valuable for showing morphology and topography on samples where backscattered electrons are most valuable for illustrating contrasts in composition in multiphase samples. The typical voltage range for SEM is 2 to 50 kV and the diameter of the beam can be varied from 5 nm to 2 μm. For SEM sample preparation, a sample is required to be completely dry for which the sample chamber is at high vacuum. To image the biological sample such as living cells under SEM, chemical fixation is required to preserve and stabilize their structure. For fixing the cell structure, incubate the cells in some fixatives such as glutaraldehyde, formaldehyde etc which causes intermolecular and crosslinks between macromolecules and provide ultra-structural preservation. We have used the SEM [(ZEISS, Germany)- EVO-18 special edition] to study the surface morphology of human cancer cells. Fig.2.6.(a) shows the schematic diagram of scanning electron microscope and (b) represents the SEM image of A549 cell.

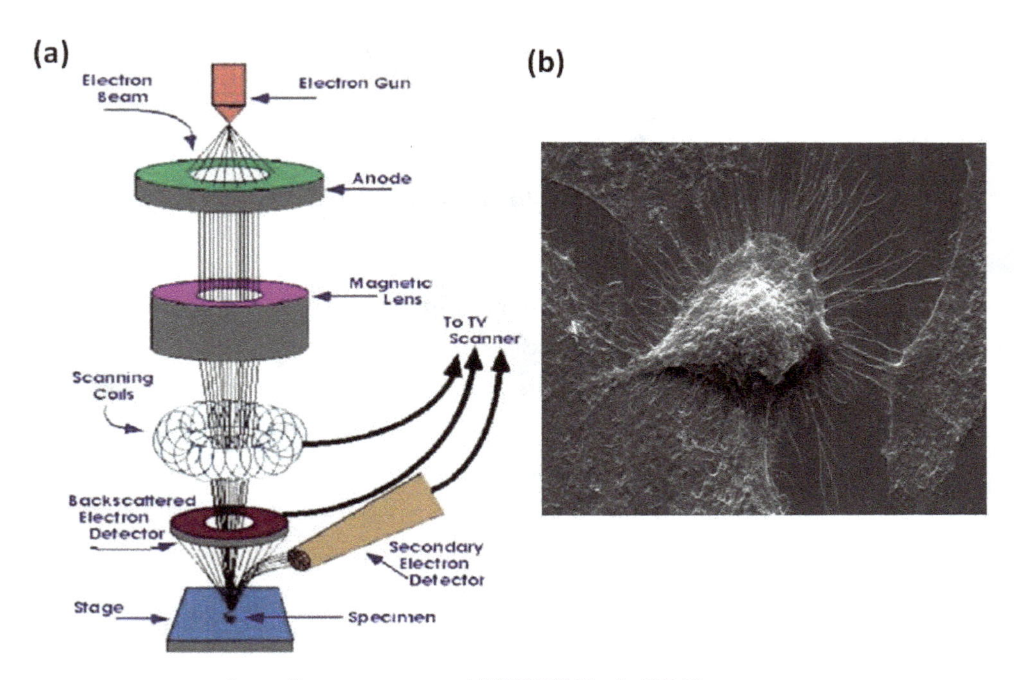

https://images.app.goo.gl/4533BNVExt1pfQRJ6

Fig.2.6. (a) Schematic diagram of scanning electron microscope, (b) SEM image of lung cancer cell (A549).

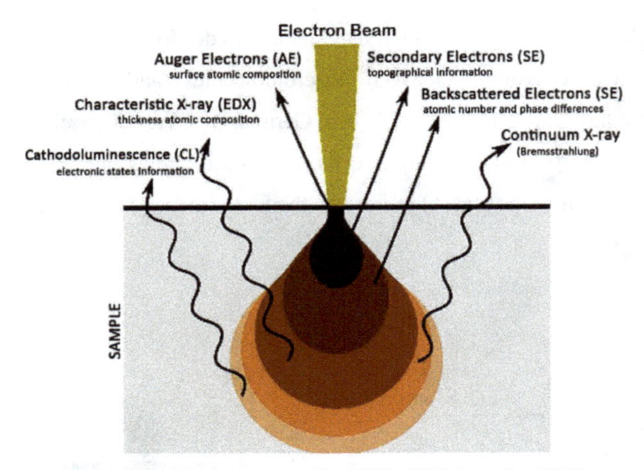

https://images.app.goo.gl/RmWBxT7L4mepiwAD6

Fig.2.7. Schematic representation of different signals in SEM that comes out from sample.

2.4.2.2. Transmission electron microscopy:

In transmission electron microscopy technique, a beam of electron is transmitted through a thin specimen and image is formed. This image formation occurs by interacting of electrons with the sample. For TEM imaging, the specimen should an ultrathin section less than 100nm thick or a suspension on a grid. Fig.2.8.(a) showing schematic diagram of a TEM instrument. A typical TEM contains the following components: electron gun, a vacuum system where the electrons travel, electron column for generation of the electron stream, a series of electromagnetic lenses, detectors, water chilling system, specimen chamber and imaging devices. The electron gun generates the electron beam and the beam travel through electron column. The electron column is made up of the gun assembly at the top, a column filled with a set of electromagnetic lenses, the sample port and airlock and a set of apertures which can be moved in and out of the path of the beam. The electron lenses are designed to act in a manner which focused the parallel electrons at some constant focal distance. In TEM, the majority of electron lenses operate electromagnetically. The most common detectors seen on a TEM is the x-ray energy dispersive spectroscopy (EDS or EDAX) system. The electron beam strikes the specimen and part of it are transmitted which is focused by the objective lens into an image. In TEM, absorption of electrons plays a very minor role in image formation. The TEM contrast depends on deflection of electrons from their primary transmission direction when they pass through the specimen. The contrast is generated when there is a difference in the number of electrons being scattered away from the transmitted beam. The electron beam pass through Bragg scattering by following the Bragg's law ($2d \sin\theta = n\lambda$). We have used TEM (JEM-2100HR-TEM, JEOL, Japan) to obtained the microstructure of synthesized nanoparticles. Fig.2.8.(b) is the TEM image of cobalt ferrite nanoparticle. The lattice fringe of the nanoparticles was obtained on high resolution TEM (HRTEM) mode where the resolution is about 0.2nm.

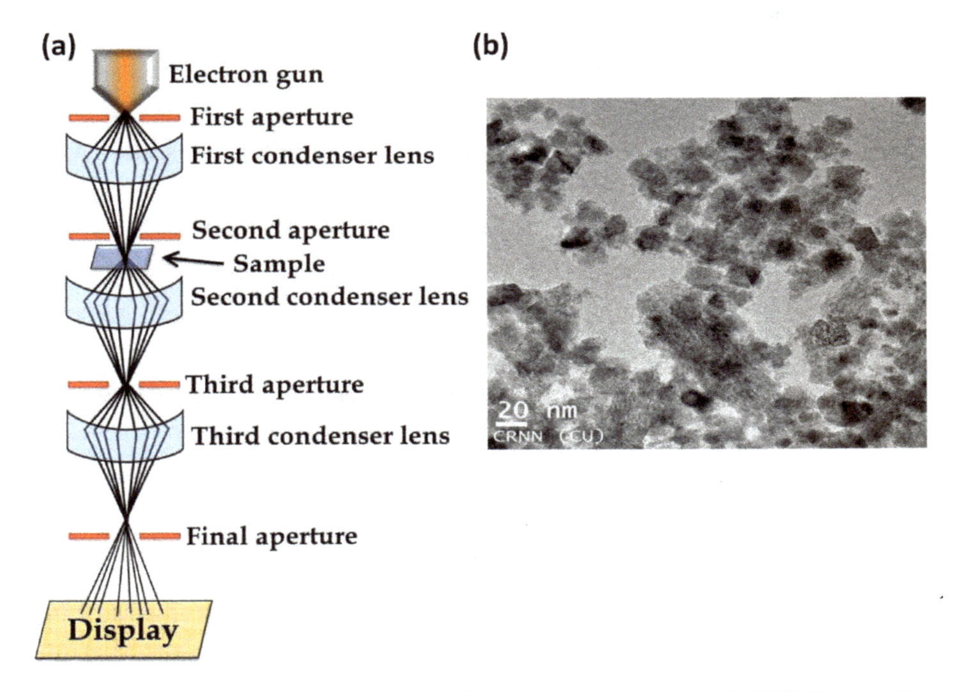

Fig.2.8. (a) Schematic diagram of transmission electron microscopy, (b) TEM image of $CoFe_2O_4$ nanoparticle.

2.5. Optical characterization techniques:

2.5.1. UV-visible absorption spectrometer:

Spectroscopy is the measurement and interpretation of electromagnetic radiation which is absorbed or emitted by the molecules or atoms or ions of a sample when they are moves from one energy state to another energy state. Absorption spectrophotometry is the measurement of absorption of radiant energy by various substances which involves the measurement of absorptive capacity for radiant energy in the visible, UV and IR regions of the spectrum. In visible range (λ=400-800nm), the coloured sample absorbs light of different wavelength and get an absorption curve. In UV spectroscopy, the UV radiation is absorbs by the sample in ranges from 200-400nm. Valence electrons of the sample absorb the energy which causes transition of molecules from ground state to excited state. This absorption is characteristic and depends on the nature of electron present. The Beer-Lambert law is used to measure the concentration of absorbing species in the sample solution is given below.

$$A = \log_{10}\left(\frac{I_0}{I}\right) = \varepsilon.c.L$$

where, A is the measured absorbance of the sample, I_0 is the intensity of the incident light at a given wavelength, I is the transmitted intensity of light, L is the path length of light through the sample, c is the concentration of the absorbing species and ε is the molar extinction coefficient that is constant for a particular absorber. According to the Beer's law, the intensity of beam of monochromatic light passed through a solution decreases exponentially with increase in concentration of absorbing species. On other side, Lambert's law states that the decrease rate of intensity of monochromatic light with the thickness of the medium is directly proportional to the intensity of incident light. The schematic diagram of the UV-visible spectrometer is showing on Fig. 2.9. We take the UV-visible absorbance spectra of our samples using Shimadzu UV-2600 spectrophotometer.

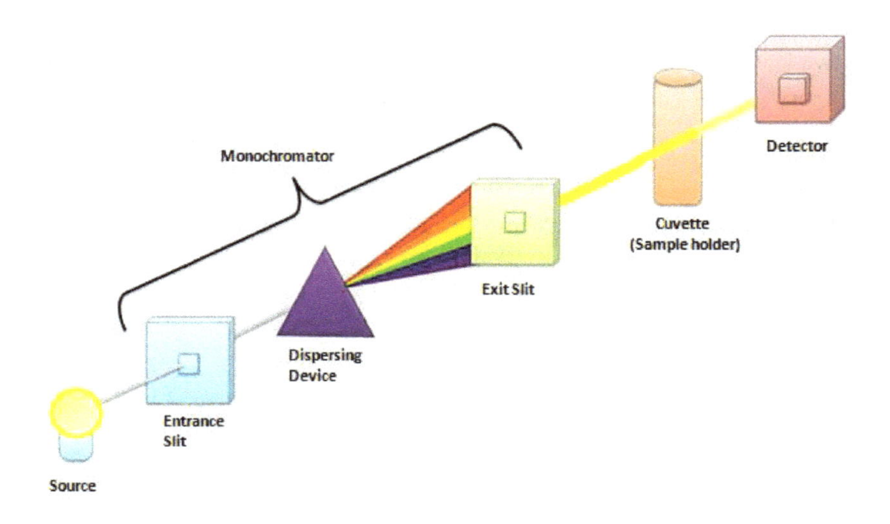

https://images.app.goo.gl/g7bTr63CvSpF3Y7P8

Fig.2.9. Schematic diagram of UV-visible absorption spectrometer.

2.5.2. Fourier Transformed Infrared Spectroscopy (FTIR):

Fourier-transform infrared spectroscopy (FTIR) is a technique by which an infrared spectrum of absorption or emission of a solid, liquid or gas is obtained [11]. An FTIR spectrometer simultaneously collects high-resolution spectral data of the sample over a wide spectral range. Infrared radiation stay between the visible and microwave portions of the electromagnetic spectrum and the infrared waves have wavelengths which is longer than visible and shorter than microwaves. By using this spectroscopy, the types of chemical bonds present in a molecule are identified. The molecules of the sample are bombarded with infrared radiation. When the frequency of the infrared radiation matches with the natural frequency of the bond of a molecule, the amplitude of the vibration increases and causes absorption of infrared. In FTIR spectroscopy, some of the infrared radiation is absorbed by the sample and some of it is passed through (transmitted). Hence, the spectrum represents the molecular absorption and transmission, creating a molecular fingerprint of the sample. Schematic representation of FTIR instrument is represented in Fig.2.10. The Michelson interferometer is the main key of the FTIR spectrometer which split one beam of light into two beams. The parallel IR beam is partly reflected and transmitted across the beam splitter and transfer to the stationary and moving mirrors respectively. Then the transmitted beam from the fixed mirror and reflected beam from the moving mirror allow constructive and destructive interference at the behind of the beam splitter depending on the wavelength of lights and the optical path difference which is induced by the moving mirror. This resulting signal is entitled as interferogram. Finally, Fourier transformation of this interferogram is carried out and gets a frequency spectrum. For FTIR measurements of our magnetic materials, the powdered samples were mixed with KBr powder and pelletized in a hydraulic press.

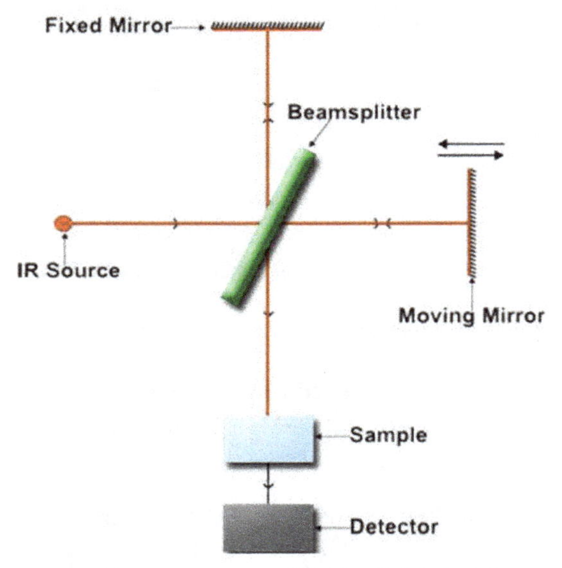

https://images.app.goo.gl/VieFcw7z5VgEzek16

Fig.2.10. Schematic diagram of FTIR.

2.5.3. Fluorescence Microscope:

A fluorescence microscope is an optical microscope which uses fluorescence to study the properties of organic and inorganic substances of sample [12,13] and generate an image. A typical fluorescence microscope consists of following components: a light source (Xenon arc lamp or mercury vapour lamp or high power LEDs and lasers), the excitation filter, the dichroic mirror and the emission filter. In fluorescence microscope, the specimen is illuminated with a specific wavelength of light which is absorbed by the fluorophores (a fluorescent chemical compound) of the sample. The fluorophores successively emits lower energy light of longer wavelengths. The spectral emission filter separates the illuminated light from the weaker emitted fluorescence. The excitation light reflects off the surface of the dichroic mirror into the sample and the fluorescence emission from the specimen again passes through the dichroic to the detecting system. Schematic diagram of a fluorescence microscope is shown in fig.2.11. We use the EVOS-FL microscope (Invitrogen, USA) for imaging of RITC tagged MNPs treated A549 cells (Fig.2.12.) Different fluorescent dyes have been used for stating of cytoplasm and nucleus.

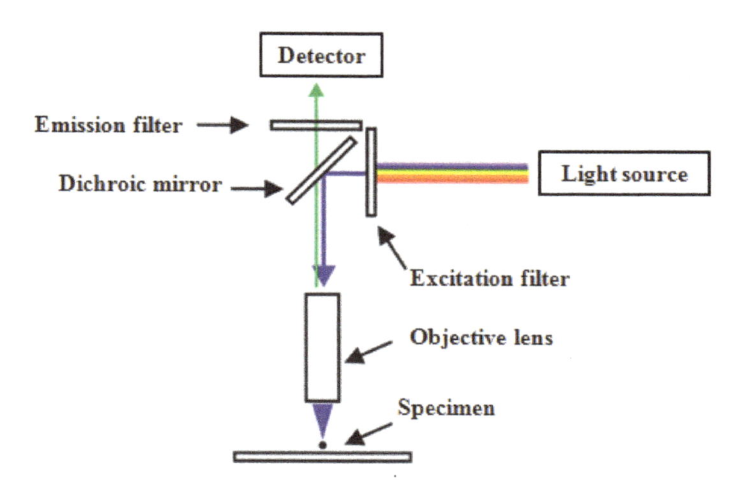

Fig.2.11. Schematic diagram of a fluorescence microscope.

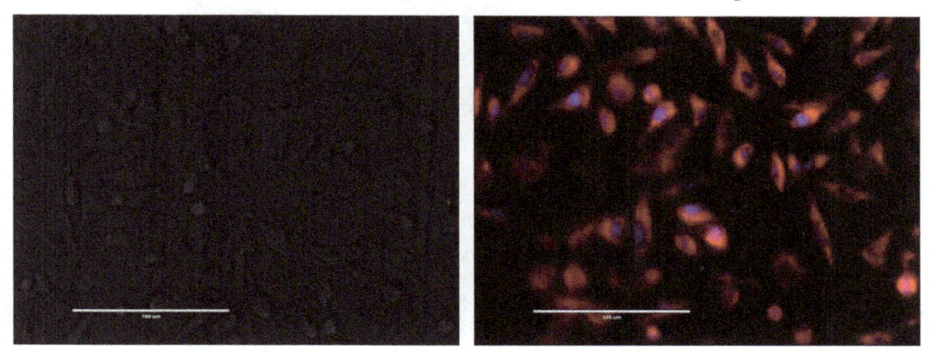

Fig.2.12. Transmittance and fluorescence image of RITC and DAPI tagged MNPs treated A549 cells seen by fluorescence microscope.

2.5.4. Confocal microscopy:

Confocal microscope is also an optical fluorescence microscope which is more complicated than the normal fluorescence microscope and gives better resolution of the fluorescence image & contrast of a micrograph. In confocal microscopy, two pinholes are used. A pinholes is placed in front of the illumination source which transmit laser light through a small area and imaged onto the focal plane of the specimen at once instead of illuminating the whole sample. Hence, the fluorescent light is emitted from exactly the focal point of the specimen. Another pinhole is placed right in front of the detector which cuts off signals that are comes from out of the focus and allows only the fluorescence signals from the illuminated focal plane to enter the light detector. The schematic diagram of a confocal microscopy is shown in Fig.2.13. Herein, fluorescence micrographs of MDAMB-231 cell treated with RITC tagged MNPs (Fig.2.14.) were taken using a cofocal microscope (Olympus Model IX81).

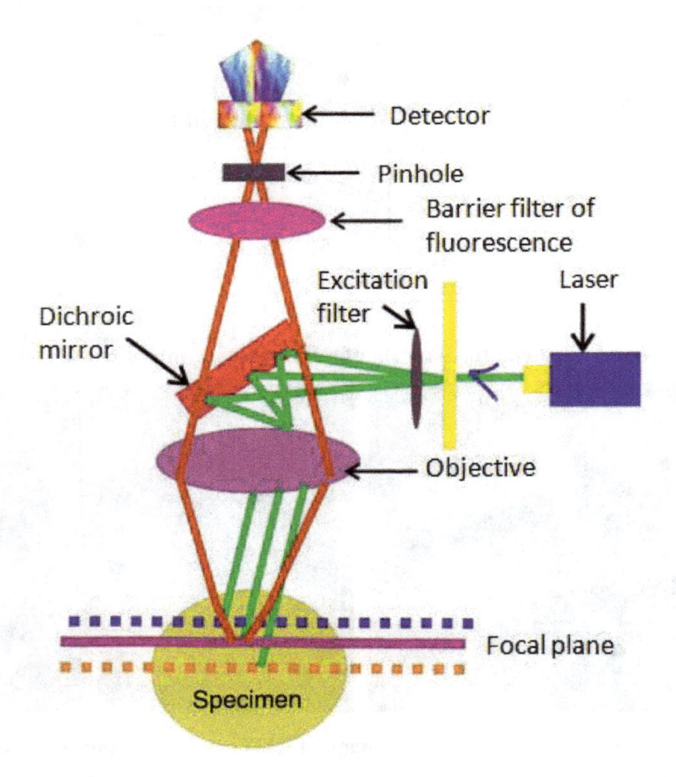

https://images.app.goo.gl/HdSxenbgiPJSwxdK8

Fig.2.13. Schematic diagram of a confocal microscope.

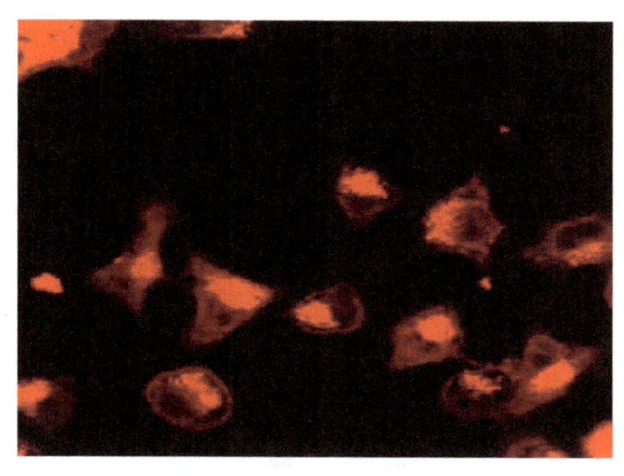

Fig.2.14. Confocal image of RITC tagged MNP treated MDAMB-231 cell.

2.6. Isothermal titration calorimetry (ITC):

Isothermal titration calorimetry (ITC) is a physical technique that used to measure the heat discharged or consumed all along a bimolecular reaction. It is an analytical method by which interaction between a small molecules and a larger macromolecules can measure directly and the interaction results in either heat generation or absorption depending on the type of binding which is exothermic or endothermic [14]. ITC consists of two cells: sample cell and reference cell shown in Fig.2.15. Macromolecule is stored in the sample cell chamber where the small molecule is injected into that cell from a syringe. The reference cell contains only the buffer solution. A steady temperature and pressure is maintained in both cells. The interaction into the sample cells release heat and the instrument can determine the amount of heat energy released. Both the reference cell and sample cell have very precise thermometers which help to measure the heat change. By this reaction the binding affinity (K_a), enthalpy changes (ΔH), Gibbs free energy changes (ΔG) and entropy changes (ΔS) can be determined by using the following relationship:

$$\Delta G = -RTInK_a = \Delta H - T\Delta S$$

Where, R is the gas constant and T is the absolute temperature.

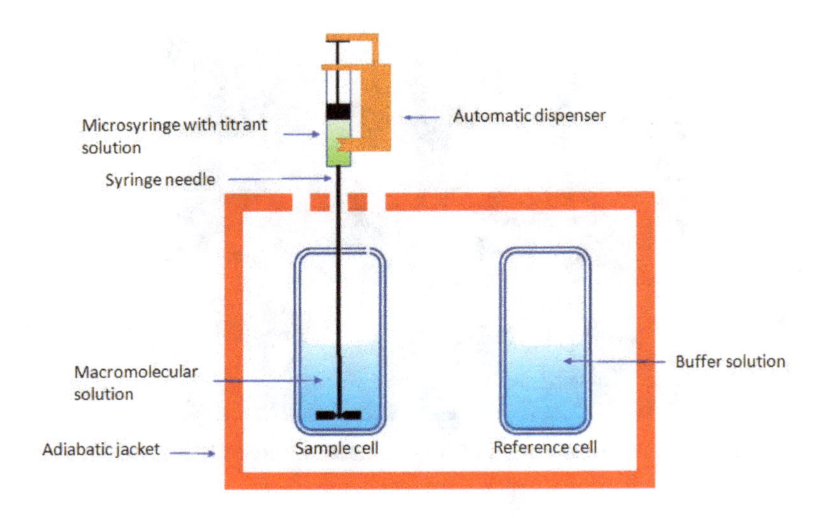

https://images.app.goo.gl/q5beuXTqXN2GsciQ6

Fig.2.15. Schematic reprentation of ITC.

2.7. Magnetic measurements

2.7.1. Vibrating Sample Magnetometer (VSM):

Vibrating sample magnetometer is an instrument by which the magnetic properties of a sample can measures. Fig.2.16. shows schematic diagram of VSM. An electromagnet is used in VSM for applying field to the sample. In VSM, two identical wound pickup coils are placed oppositely close to the sample location. The sample is placed in between the two pickup coils which oscillated the sample sinusoidally in a given frequency. The sample is magnetized by an applied DC field whose mechanical vibration causes changes in magnetic flux density which links with the pickup coils. According to the Faraday's law:

$$V = -NA\frac{dB}{dt}$$

Where, V is the induced voltage in the pickup coils, N is the turns number in the pickup coil, A is its cross sectional area, B is the magnetic flux density and t is the time. If the magnetization of the sample is to be M then the change in magnetic flux density can be written as,

$$\Delta B = \mu_0 M$$

The signal becomes:

$$Vdt = -\mu_0 NAM$$

For measuring of the AC and DC magnetic properties of the sample, the basic principle is very similar. In AC measurement, sample shows an oscillating response caused due to varying of applied field which induce signal to the secondary coils. In case of DC measurement, a spatial and temporal oscillation flux is generated by the vibrated sample due to applied of DC field. Yet, the induced signal is depends on the frequency and amplitude of the sample's vibration and proportional to the magnitude of the sample's moment but this can add an error. This problem is resolving by a vibrating capacitor which induce the reference signal that change in the similar way as the pickup coils signal with the vibrational frequency, moment and amplitude of the sample.

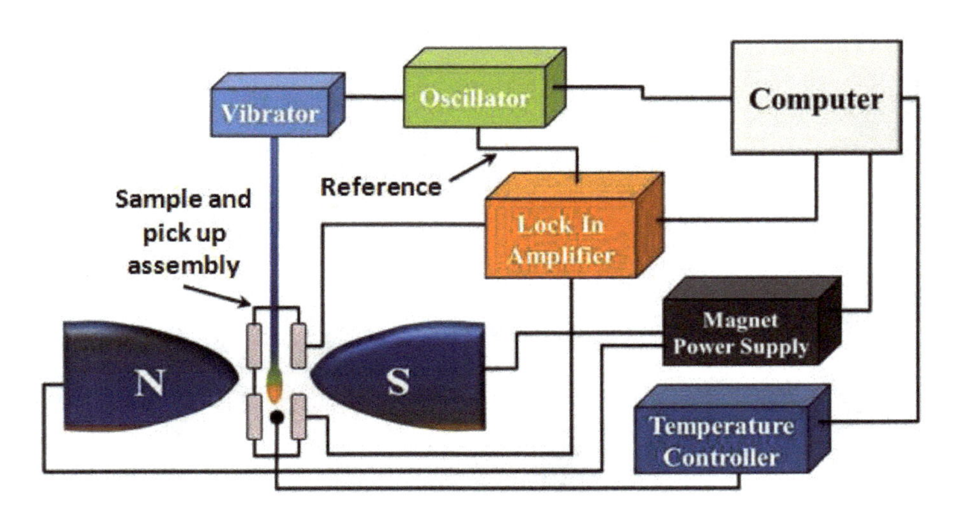

Fig.2.16. Schematic diagram of vibrating sample magnetometer.

2.7.2. Superconducting Quantum Interference Device (SQUID):

The SQUID is a very sensitive and effective magnetometer which can detect detects extremely small magnetic field and moment depending on superconducting loops containing Josephson junctions. Fig.2.17. shows the two parallel Josephson junctions in a DC SQUID which is generated by separation of two superconductors by two thin insulating layers. Without any external magnetic field, the applied biasing current I split into the two branches equally. If a small external magnetic field is applied to a superconducting loop, a screening current I_S is induced which cancel out the externally applied field. The generated I_S has the same direction as I in one of the branches of the superconducting loop ($I/2 + I_S$), and is opposite to I in the other ($I/2$ - I_S). Across the junction, a voltage is generated when the current in either branch overcomes the critical current (I_C) of the Josephson junction. Because the magnetic flux enclosed by the superconducting loop, so it must be an integral number of flux quanta (Φ_0). When the external flux is increased beyond $\Phi_0/2$ without screening the flux, then it energetically prefers to increase the external flux to Φ_0. So, the screening current turn over its direction every time the flux is changed by $\Phi_0/2$. The SQUID works in the resistive mode if the biasing current is higher than I_C. Hence, the voltage changes as a function of the applied magnetic field and the period ~ Φ_0. So, in DC SQUID the characteristic of I-V is hysteretic, the hysteresis is eliminated by connecting of a shunt resistance with the junction. We use SQUID magnetometer (Quantum Design MPMS squid VSM) to observe magnetic properties of our samples between room temperature (300K) and 5K.

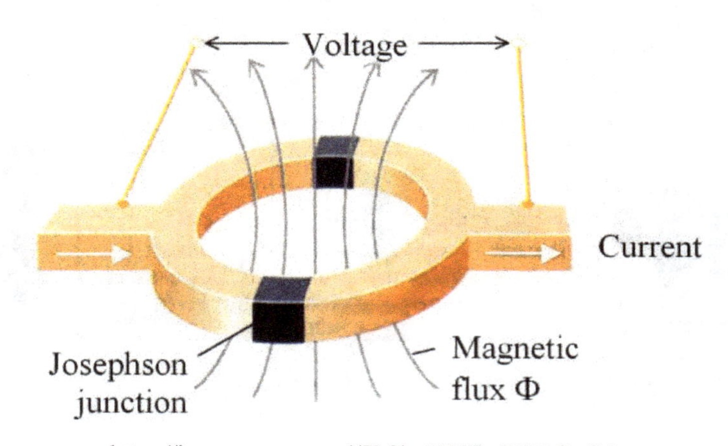

https://images.app.goo.gl/7L3bp1PNTmDDDQwE6

Fig.2.17. Schematic representation of SQUID showing Josephson junction.

2.7.3. AC magnetic measurement system:

Under an AC magnetic field, a magnetic material becomes oscillated which can be measured by a pickup coil. An AC hysteresis measurement setup is build with a combination of the primary and secondary circuits is shown in Fig.2.18. The main components of the primary circuit are a function generator, resistor, power amplifier, variable capacitor and primary coil. In primary circuit, a sinusoidal voltage is applied from the function generator to the input of power amplifier where the voltage get amplified and brought up to the primary coil. The coil containing current can measured from the voltage of the series resistor that is proportional to the applied AC magnetic field. The secondary circuit build up with a pickup coil pair which is compensated by connecting them in series opposition that causes zero signals in absence of sample and integrator. But, when a sample is placed in one of the pickup coils, the secondary coils provide non zero signals which is corresponds to the magnetic induction of the sample. We have performed the AC magnetic measurements in our laboratory setup instrument where we can produce about maximum 90kA/m AC magnetic field with 50 to 700 Hz frequency range. Fig.2.19. shows the AC magnetic hysteresis loops of magnetite nanoparticle.

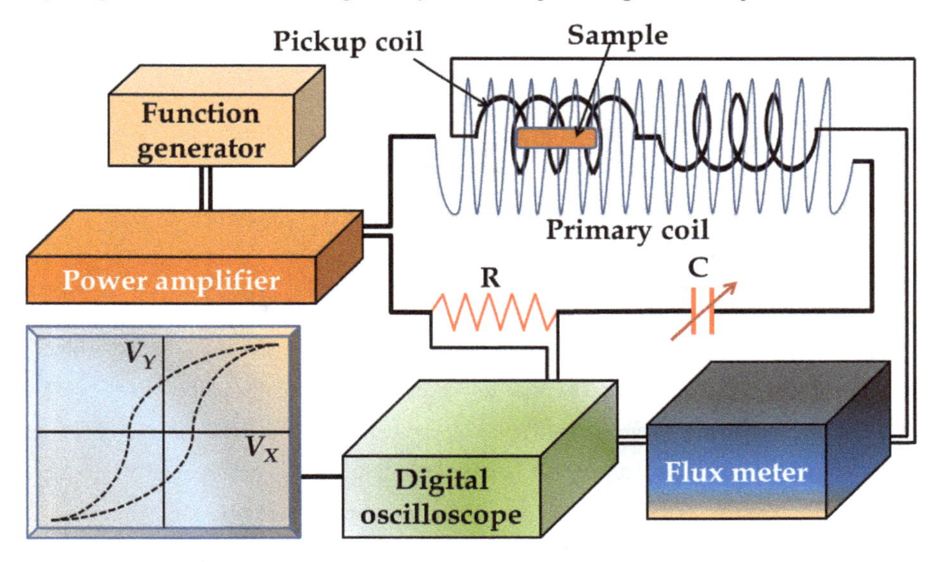

Fig.2.18. Schematic diagram of the hysteresis loop measurement by AC magnetic setup.

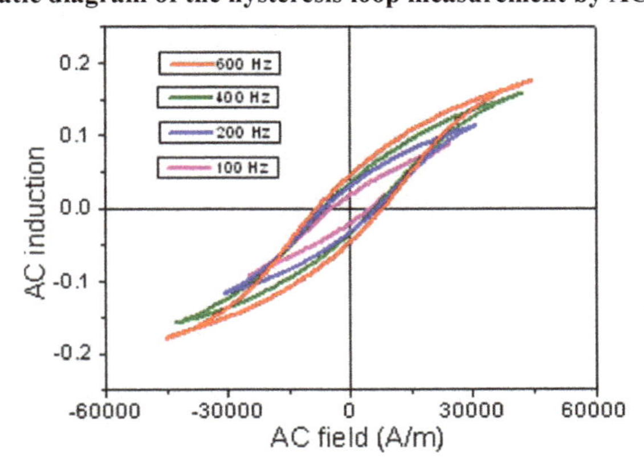

Fig.2.19. Measurement of Hysteresis loops of magnetite nanoparticle at different AC field frequencies.

2.8. CO_2 Incubator:

A cell culture incubator is a device which is designed to maintain optimal temperature, humidity for the growth of culture cells under CO_2 atmosphere. In an incubator the temperature sets from 4^0C to 50^0C and CO_2 concentration run from 0.3 to 19.9%. The interior part of incubator is basically made by non-corrosive stainless steel but some incubator also made with antimicrobial copper surfaces to prevent contamination. In CO_2 incubator, temperature is controlled either by a water bath that circulates through the walls of the cabinet called water jacketed CO_2 incubator, or by electric coils which give radiant heat. The humidity of incubator is maintained between 95% to 98% by an atomizer system or water reservoir. We have maintained the human cancer cells and normal PBMC cells in an incubator by maintain 37^0C and 5% CO_2.

2.9. Laminar flow cabinet:

A laminar flow hood is an enclosed cabinet which is used to prevent contamination of biological samples, semiconductor wafers etc from the environment. Generally, the cabinet is made up of stainless steel with no gaps or joints to prevent the collection of spores [15]. A glass shield is present in front of the cabinet to protect the hood from users. A filter pad HEPA (High-efficiency particulate air) is present within the cabinet through which air is passes into the cabinet that makes the environment more sterile for work. This filter traps the dust particles, microbes etc which are enters from the working environment into the cabinet. Below the filter, a fan is present that sucks in the air and moves it around the cabinet. Laminar flow cabinets also have a UV germicidal lamp which sterilizes the interior of the cabinet and essential equipments before usage to prevent contamination of the experiment. Such of the laminar flow hoods exist in horizontal configuration and some are in vertical. I have used the laminar flow hood for sterilized cell culture experiment.

2.10. Centrifuge:

A centrifuge is a device which uses centrifugal force for separation of components of a mixture on the basis of their size, density, the viscosity of the medium, and the rotor speed. In centrifuge, the denser molecules move towards the periphery of the centrifuged tube while the less dense particles move to the centre. In a laboratory centrifuged, the radial acceleration causes denser particles sink down i the bottom of the sample tube, while low density substances are on the top [16]. Relative centrifugal force (RCF) is the measure of the rotor strength. RCF is the perpendicular force that acts on the sample in the tube which is relative to the gravity of the earth. The formula of calculate the RCF is:

$$\text{RCF (g Force)} = 1.118 \times 10^{-5} \times r \times (\text{RPM})^2$$

where r is the radius of the rotor (in 61entimetres), and RPM is the speed of the rotor in rotation per minute.

2.11. Magnetic stirrer:

In a magnetic stirrer, a rotating magnet or a stationary electromagnet is present which creates a rotating magnetic field. In laboratory magnetic stirrer is generally used for proper mixing of a solution using a magnetic stir bar. A magnetic stirrer also consists of heating system for heating the solution (Fig.2.20.). In magnetic stirrer, stir bar are used to agitate a liquid sample for proper mixing. Generally, the stir bars are coated with Teflon which is chemically inert and do not react with the solution. I have used the magnetic stirrer for synthesis of nanoparticles, drug loading, functionalization of MNPs and some other purposes.

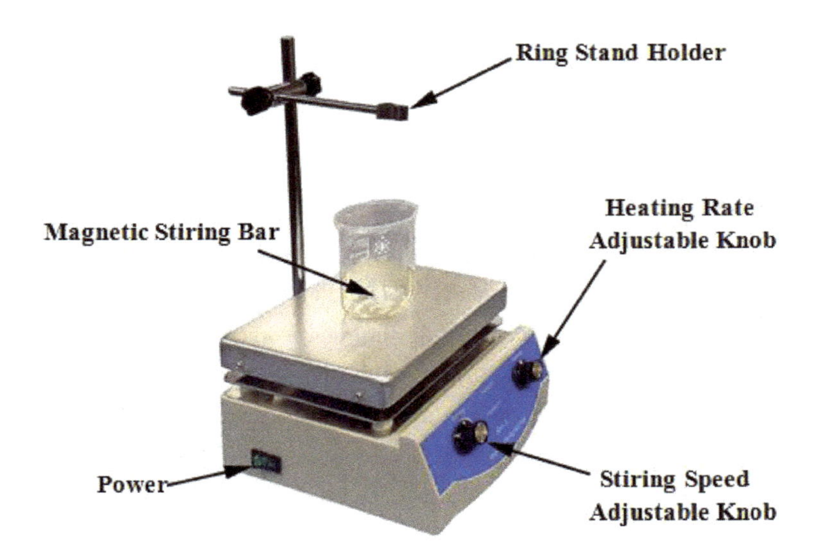

https://images.app.goo.gl/vURQ3asbdccJjTW99

Fig.2.20. Diagram of a magnetic stirrer.

2.12. Sonicator:

Sonication is the process where sound energy is applied to agitate the particles in a sample. The sonication also known as ultrasonication due to used of ultrasonic frequencies (>20 kHz) in this instrument. In my experiment, ultrasonic bath and ultrasonic probe have used for different purposes.

2.13. Microplate reader:

It is an instrument by which we can measure the chemical, biological or physical reactions of a sample within microtiter plates. This plate consists of small wells in which reaction of the sample take place. The samples have been pipette into the microtiter plate and a microplate reader detects light signals which are produced by the samples. The optical properties of these samples are the result of these reactions. Absorbance, fluorescence intensity, luminescence, time-resolved fluorescence etc are common detection modes of the microplate reader in laboratories. In microplate reader, such light signals are produced by the samples that are measured by a detector generally a photomultiplier tube (PMT). The PMT then convert these signals photons into electricity and quantified by the reader. In this measuring procedure, samples may need to be excited by light at specific wavelengths. To confirm the specific excitation of the sample, there is specific excitation filter or monochromator present in this instrument. We have used the Bio-Rad (model 550) microplate reader to determine MTT cell viability assay.

2.14. Flow cytometry:

Flow cytometry is a laser based technology by which physical and chemical characteristics of cells and particles can detect and measure [17-20]. In flow cytometry, the cells or particles are suspended in a fluid that is injected into this instrument where the sample is focused to flow one cell at a time through a laser beam. By this instrument, the cells or particles are analysed for visible light scatter and one or more fluorescence parameters. The visible light scatter is measured in the forward direction called forward scatter (FSC) and at 90° called side scatter (SSC). For light scatter, samples are labelled with a fluorescent marker that gets activated under a focused light source and emits light which is filtered and detected by light detector. Different fluorophores are used such as FITC, propidium Iodide etc. Schematic diagram of a flow cytometer is shown in Fig.2.21.

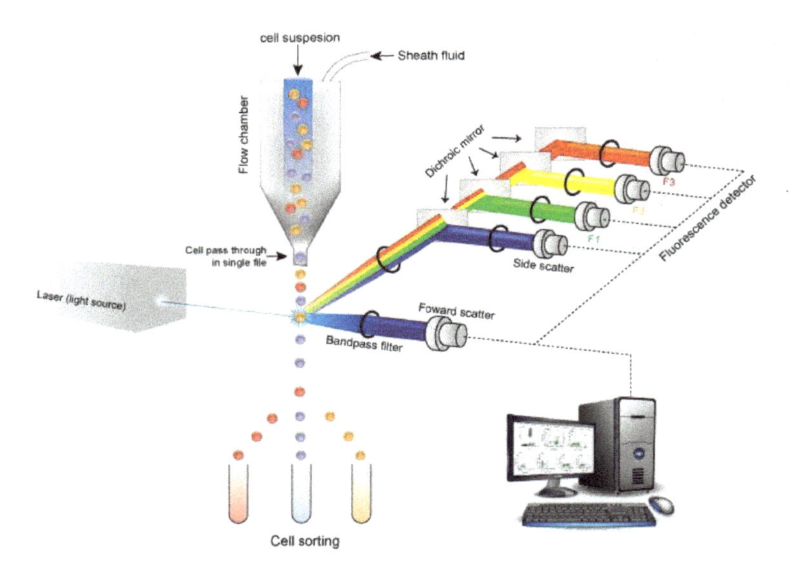

https://images.app.goo.gl/cq8jVQLidLYbqmqUA

Fig.2.21. Schematic diagram of flow cytometer instrument.

2.15. Hyperthermia circuit:

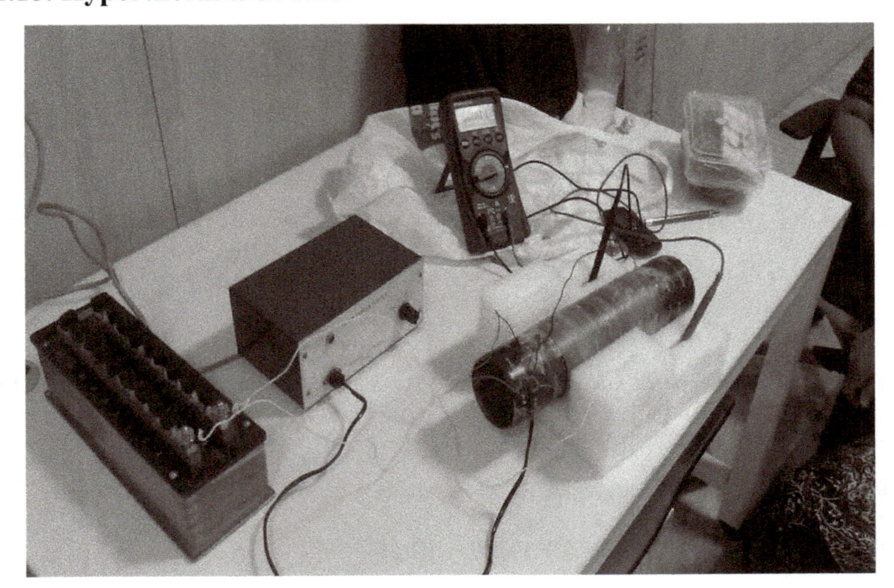

Fig.2.22. Hyperthermia circuit

Bibliography:

[1] Shi, W., Song, S., Zhang, H., 2013. Hydrothermal synthetic strategies of inorganic semiconducting nanostructures. Chem. Soc. Rev., 42, 5714-5743.

[2] Gao, S., Lin, Y., Jiao, X., Sun, Y., Luo, Q., Zhang, W., et al., 2016. Partially oxidized atomic cobalt layers for carbon dioxide electroreduction to liquid fuel. Nature, 529, 68–71.

[3] Sun, Y., Liu, Q., Gao, S., Cheng, H., Lei, F., Sun, Z., et al., 2013. Pits confined in ultrathin cerium (IV) oxide for studying catalytic centers in carbon monoxide oxidation. Nat. Commun., 4 (1), 2899.

[4] Sun, Z., Liao, T., Dou, Y., Hwang, S.M., Park, M.S., Jiang, L., et al., 2014. Generalized selfassembly of scalable two-dimensional transition metal oxide nanosheets. Nat. Commun., 5 (1), 3813.

[5] Yoo, D., Kim, M., Jeong, S., Han, J., Cheon, J., 2014. Chemical synthetic strategy for single-layer transition-metal chalcogenides. J. Am. Chem. Soc., 136 (42), 14670-14673.

[6] Walton, R.I., 2002. Subcritical solvothermal synthesis of condensed inorganic materials. Chem. Soc. Rev., 31, 230–238.

[7] Kumar, S., Nann, T., 2006. Shape Control of II–VI Semiconductor Nanomaterials. Small, 2 (3), 316–329.

[8] Penn, R.L., Banfield, J.F., 1998. Oriented attachment and growth, twinning, polytypism, and formation of metastable phases; insights from nanocrystalline TiO 2 Am. Mineral., 83, 1077-1082.

[9] Tang, Z., Kotov, N.A., Giersig, M., 2002. Spontaneous Organization of Single CdTe Nanoparticles into Luminescent Nanowires. Science, 297, 237–240.

[10] Burda, C., Chen, X., Narayanan, R., El-Sayed, M.A., 2005. Chemistry and Properties of Nanocrystals of Different Shapes. Chem. Rev., 105, 1025–1102.

[11] Griffiths, P., De Hasseth, J.A., Winefordner, J.D.,2007. Fourier Transform Infrared Spectrometry, 2nd Edition. Wiley-Blackwell, ISBN 978-0-471-19404-0.

[12] Spring, K.R., Davidson, M.W., 2008. Introduction to Fluorescence Microscopy. Nikon MicroscopyU. Retrieved 2008-09-28.

[13] Microscopes—Help Scientists Explore Hidden Worlds. The Nobel Foundation. Retrieved 2008-09-28.

[14] Indvk, L., Fisher, H.F., 1998. Theoretical aspects of isothermal titration calorimetry. Methods Enzymol, 295, 350-364.

[15] "Types of Laminar Flow Cabinets – Uses and Benefits – Information Guide". www.laminarflows.co.uk. Retrieved 19 April 2018.

[16] Mikkelsen, S.R., Cortón, E., 2004. Bioanalytical Chemistry, Ch. 13. Centrifugation Methods. John Wiley & Sons, Mar 4, 247–267.

[17] Picot, J., Guerin, C.L., Le Van Kim, C., Boulanger, C.M., March 2012. "Flow cytometry: retrospective, fundamentals and recent instrumentation". Cytotechnology, 64 (2), 109–130.

[18] "flow cytometry". TheFreeDictionary.com. Retrieved 2018-09-18.

[19] "Practical Flow Cytometry - Beckman Coulter". www.beckman.com. Retrieved 2018-09-18.

[20] Givan, A.L., 2011. "Flow Cytometry: An Introduction". Methods in Molecular Biology, 699, 1–29.

Chapter 3

Use of magnetic nanoaprticles for hyperthermia therapy and cell imaging

In this chapter we have discussed about the synthesis of single and multi-domain magnetic nanoparticles (MNPs) and their surface modification. Here we have also focused on characterization of these MNPs and also on the magnetic measurements under AC and DC field. We have studied their possible applications in biomedical field such as cell imaging, in-vitro cytotoxicity study and hyperthermia applications etc. We have also discussed the extraction of organic dye and tagged this dye with MNPs for imaging of multiple human cells.

3.1. Induction:

In recent years, researchers give an immense attention on therapeutic and diagnostic uses of nanoparticles in various fields of medical sciences such as for bio imaging [1], drug delivery [2,3], hyperthermia [4,5], biomedicine [6], gene delivery [7], targeted therapy [8,9], bio separation [10], organ repair [11] etc. due to their high surface to volume ratio, high stability, and high porosity which make them suitable for proper use.Until now cancer treatments are incapable to cure a patient completely and also have lot of harmful and fatal side effects [12]. Recently, the magnetic nanoparticles take an attention in researchers mind due to its in situ monitoring capability under external magnetic field. Hyperthermia is a heating method which has been introduced in medical field for cancer treatment as an alternative method which have less side effects compare to other heating treatments. Hyperthermia treatment is based on rising of temperature to the range of 42-45^{0}C in cancerous cells that causes cancer cell necrosis without harming the healthy tissues [13-15].Cancer cells faces poor blood flowbecause of their uncontrolled growth, which causes lack of oxygen in tumour region and makes it more acidic that results cancer cells more sensitive to heat[16]. Magnetic hyperthermia is a therapy to treat cancer where magnetic nanoparticles are used which generates heat under AC magnetic field. This method involves two stages, at first; the MNPs are needed to deliver into the tumour site and after that placing the tumour under an AC induced magnetic field [17-19]. But use of such magnetic nanoparticles in cancer treatment have some limitations such as poor heating efficiency, biocompatibility, internalization of particles inside the cell, stability etc [20]. Therefore, we should give major efforts on synthesis of such MNPs by tuning some key parameters like size, shape, surface modification of the nanoparticles, internalization of the particles with cell, magnetic anisotropy etc. in order to make the MNPs with optimizedheating efficiency, biocompatibility, chemical stability, uniform dispersion in liquid medium etc. The most important magnetic property of the MNPs is that its coercivity (H_C) which is changed depending on the NPs size. When the particle size decreases below a critical diameter (DS), they become single domain and the coercivity reaches maximum in the single domain particles. But further decrease of the particles size causes decrease of the H_C and finally reaches to zero. In that case the particles become superparamagnetic. Under AC magnetic field, the superparamagnetic or ferromagnetic NPs are suitable for magnetic hyperthermia therapy because of their hysteresis loss and eddy current [21].

According to previous reports, functionalization of nanoparticles with DNA [22], folic acid [23], biocompatible polymer [24], citric acid [25] etc. increases the potentiality of nanomaterials for biomedical application. Until now, many kinds of MNPs have been used in magnetic hyperthermia therapy such as MFe_2O_4 where M= Fe [26-28], Co [29-31], Mn [32], Mg [33], Mn/Mg [34], Zn/Fe [35], Fe/Mn [36] as well as some alloys as like FeCo [37], CuNi [38] etc. In addition to hyperthermia application, if the MNPs have fluorescence properties, it helps to study the internalization property of MNPs in the cells by visualizing under microscope. Generally, different fluorescent dyes have been used to stain the cells for investigation under fluorescent microscope. However, fluorescence intensity of cells has been induced when it attached with biomolecules like protein, lipid, DNA etc. Therefore, cancer cell imaging is one of the important studies for diagnosis of cancer and cures this disease in its early stage. $CoFe_2O_4$ is one of the good MNP that exhibit high magneto-crystalline anisotropy and more efficient SAR (Specific absorption rate) value that variably increases according to its size up to a certain limit [39,40]. $CoFe_2O_4$ MNPs also have a special advantage because of their stability is quite higher due to its Co which one is in +2 state and Fe in +3 state which generally reduces the aerial oxidation in such materials. So, $CoFe_2O_4$ nanoparticles (NPs) with proper size, shape and surface modification may be useful for biomedical application.

In our study, we have focused to design the magnetic nanoparticles for hyperthermia therapy which induced heat under AC magnetic field and can be applied in a controlled manner from outside of the body. Firstly, we have synthesized spherical $CoFe_2O_4$ NPs (CFMNPs) of two different sizes about 250 and 350 nm by chemical co-precipitation methods which are multi-domain in character. We have characterized those using different techniques and performed various AC and DC magnetic measurements to study their effectiveness in hyperthermia treatment.

Next, we want to see the effectiveness of the single domain $CoFe_2O_4$ NPs on hyperthermia treatment. So, we have synthesized single domain $CoFe_2O_4$ nanoparticle (CF NP) by chemical co-precipitation method and modified their surface with herring sperm DNA (CF-DNA NP) where CF NP was synthesized on DNA scaffold. A comparative study has been performed for their magnetic property and biomedical application. We have characterized these particles by using different technique and isothermal titration calorimetry (ITC) was used to confirm the binding of CF NP with DNA. The cellular uptakes of these NPs were checked after functionalization of these particles with fluorescent dye RITC. Higher uptake of the MNPs by the cancer cells would be giving the greater effect on heating. The cell viability effect of these nanoaprticles has been checked at two different temperatures i.e. normal physiological temperature (37°C) and hyperthermic temperature (~ 45°C) on MDAMB-231 cells and normal WBC (PBMC). For hyperthermia application, treated MDAMB-231 cells were kept inside a solenoid containing 174 numbers of turns and operate at 230 V with 50 Hz AC supply. The cell morphology of treated MDAMB-231 cells after hyperthermia application has been observed by using phase contrast microscope and by scanning electron microscope (SEM).

Furthermore, we have extracted water soluble dye from beet root by very easy process and used this as a fluorescent marker for imaging of different cells. We have also tagged the CF NP with this dye to observe the cellular fluorescence imaging. The cell imaging has been done on both cancer and normal cell lines such as MDAMB-231, A549, leukemia cells, squamous epithelium cells and PBMC. Here, we have checked the minimum time required to internalize this dye inside the cells.

3.2. Experimental details

3.2.1. Materials:

The following chemicals have been used for this work as mentioned below:

Iron acetyl acetonate, cobalt acetate, phenyl ether, ethylene glycol, oleic acid, urea, Ferric chloride hexahydrate ($FeCl_3.6H_2O$; purity 98%), Cobalt chloride hexahydrate ($CoCl_2.6H_2O$; purity 97%), alcohol histopaque-1077 and Rodamin B- isothiocyanate (RITC) are procured from Sigma-Aldrich. Deoxyribonucleic acid (DNA) extracted from herring sperm, potassium hydroxide (KOH, purity $\geq$ 85%), phosphate buffer solution (PBS), Dimethyl sulfoxide (DMSO), ethanols are purchased from Merck, Germany. Culture media Dulbecco's Modified Eagle's Medium (DMEM), non essential amino acids, Amphotericin B, Roswell Park Memorial Institute 1640 (RPMI-1640), penicillin, streptomycin, gentamycin, L-glutamine are procured from HIMEDIA, Mumbai, India and Fetal bovine serum (FBS) is purchased from Invitrogen, Carlsbad, CA, USA. De-ionized water (H_2O) taken from millipore system.

3.2.2. Synthesis of micelle guided cobalt ferrite ($CoFe_2O_4$) magnetic nanoparticles (CFMNPs):

For synthesis of micelle coated CFMNPs by chemical co-precipitation method, at first 3 mmol of iron acetyl acetonate and 1.5 mmol of cobalt acetate were dissolved in a mixture of 40 mL phenyl ether and 10 mL of ethylene glycol. Then, 4mL of oleic acid and 3 gm of urea was added to this solution and heated for 1 h at 160°C. Next, the samples were kept at room temperature for cooling and then collected by centrifugation followed by washing with distilled water, ethanol and acetone. After drying the particles were annealed at 300°C for 1 h. In this work, two sets of particles have been prepared by varying the oleic acid concentration only keeping all other conditions unchanged. In case of set A and set B sample, 4 mL and 2 mL of oleic acids have been used respectively.

3.2.3. Synthesis of micelle free cobalt ferrite ($CoFe_2O_4$) magnetic nanoparticles (CF NP):

We have synthesized the $CoFe_2O_4$ NP (CF NP) by chemical co-precipitation method. At first, two precursor salts $FeCl_3.6H_2O$ & $CoCl_2.6H_2O$ were taken in a conical flask with a molar ratio of 2:1 and dissolved in 100 mL of de-ionized water. To get a homogeneous mixture, the solution was stirred under magnetic stirring and after proper mixing the solution was allowed to heat. 10gm of KOH was dissolved in de-ionized water separately. When the temperature of the parent solution was attained to 85°C, the KOH solution was added drop wise to it and stirred vigorously for 1 h. Then the sample was allowed to cool at room temperature and washed with distilled water, ethanol and acetone for several times and collecting by magnetic separation. Finally, the particles were dried in air and stored in dry place. The particles were grind well before further application.

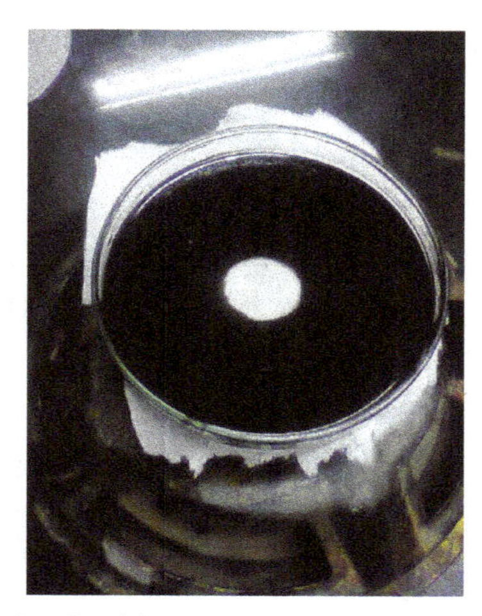

Fig.3.1. $CoFe_2O_4$ NP showing it's magnetic property when it kept on a magnet.

3.2.4. Synthesis of DNA functionalized $CoFe_2O_4$ NP (CF-DNA NP):

For synthesis of DNA functionalized CF NP (CF-DNA NP), in addition to these above solutions, another solution of DNA was prepared where 0.4 gm of DNA was dissolved in distilled water. Here, the DNA solution was mixed originally with the above mention parent solution and the rest of the protocol was same as synthesis procedure of CF NP. We also have synthesized the different batches DNA functionalized CF NP by varying the DNA amount (0.2 gm and 0.8 gm).

3.2.5. Characterization of these NPs:

The crystalline phase, size, structure of nanoparticles was analysed by X-ray diffraction pattern (XRD). The XRD data of two sets of CFMNPs, CF NP and CF-DNA NP was recorded in Rigaku Miniflex II desktop X–ray diffractometer using Cu Kα (λ= 1.5418 Å) radiation with 2θ ranging from 20°-80° at 1° per minute scanning rate (at 40KV & 40 mA). Surface morphological property and microstructure of these particles were studied by transmission electron microscopy (JEM-2100HR-TEM, JEOL, Japan) and scanning electron microscope (SEM, QUANTA FEG 250). For TEM and SEM analysis, the grinded particles were well dispersed in alcohol and drop casted on 300 mesh carbon coated copper (Cu) grid and on silicon wafer respectively and then dried in air. The surface chemistry of these particles was investigated by Fourier transform infrared (FTIR) spectroscopy (NEXUS-470, Nicolet, USA) by KBr (potassium bromide, IR grade, Sigma-Aldrich) matrix method at room temperature in the frequency range of 4000- 400 cm^{-1}. The binding affinity of CF NP with DNA was studied by isothermal titration calorimetry (ITC) (Malvern Model Microcal ITC 200). The DC magnetic properties of the CFMNPs at room temperature up to a magnetic field of 1.6 T was measured by vibrating sample magnetometer (VSM, Lake Shore Model-7144) where the AC magnetic field dependent measurements of these particles were performed in our own laboratory made AC hysteresis measurement setup. On other side, the magnetic properties of CF NP and CF-DNA NP were carried out between room temperature (300K) and 5K by SQUID magnetometer (Quantum Design MPMS squid VSM). The absorption spectra were taken by using Shimadzu model UV-2600 spectrophotometer using 1 cm path length of a quartz cuvette.

3.2.6. Hyperthermia treatment by CF NP and CF-DNA NP:

For hyperthermia application, a basic solenoid has been made in our laboratory which contains 174 no. of turns and operating at 230 V with 50 Hz AC supply shown in Fig.2.22. Inside the solenoid, around 7.897 mT magnetic field and 0.729 mH inductance has been produced by the coil.

3.2.7. Heating efficiency of CF NP under AC magnetic field:

To check the heating effect of CF NP in hyperthermic condition, we have prepared a solution of particles where 1 mg of CF NP was dissolved in 2 mL PBS solution in a 2 mL plate and then the plate was placed inside the above mention solenoid. We also kept another plate inside the solenoid containing 2 mL of PBS only. AC field was then applied for several times and the temperature produced by the particles was measured at different time interval by using a multimeter probe. The temperature of the particle dissolved in PBS solution was compared with the plate containing only PBS solution. After removing of magnetic field, we have noted the time required by the particles to cool down in its initial temperature.

3.2.8. Details of cell culture experiments using CF NP and CF-DNA NP:

3.2.8.1. Cell culture:

The human mammary carcinoma cell lines (MDAMB-231) was obtained from National Centre for Cell Science (NCCS, Pune, India) and maintained in Dulbecco's Modified Eagle's Medium (DMEM) containing 10% of fetal bovine serum (FBS) at 37^{0}C in 5% CO2 incubator. To avoid contamination, antibiotics such as penicillin, streptomycin and gentamicin (100mg/L each) were added to the DMEM.

3.2.8.2. Peripheral blood mononuclear cell (PBMC) isolation:

The human whole blood was collected in heparinised vacutainer blood collection tubes (BD, Franklin Lakes, NJ) from a healthy adult person with prior permission. After that the blood was layered at 45^0 angles in centrifuged tubes onto 120 mL of Histopaque 1077 gradient. Then the tubes were centrifuged for 30 minutes at 3000 rpm. After centrifugation, the peripheral blood mononuclear cell (PBMC) layer was collected and washed twice with phosphate buffer solution (PBS) and maintained in Roswell Park Memorial Institute (RPMI-1640) medium with 10% FBS in CO2 incubator. The cells were then treated with CF NP and CF-DNA NP to study its effect on cell viability.

3.2.8.3. Labelling of nanoparticles with RITC:

To check the internalization property of these nanoparticles inside the MDAMB-231 cells, both CF NP & CF-DNA NP were labelled with RITC. Firstly, 25 mg of CF nanoparticles was dispersed in 0.1 M $NaHCO_3$ solution. Then the solution was sonicated for proper mixing. On other side, 1 mg of RITC was dissolved in 2 mL of aqueous DMSO (1:1) and formed RITC solution. After that, this solution was mixed with the above $NaHCO_3$ made CF dispersion. Then, the new mixture was continuing stirred for 24 h at room temperature in dark condition. Next day, the RITC tagged CF NP was centrifuged at 10000 rpm and washed gently with water for removing of excess untagged RITC. To follow the same protocol, the RITC tagged CF-DNA NP was also prepared. Both RITC tagged CF NP & CF-DNA NP were dried and stored at dark place.

3.2.8.4. In vitro cell viability study:

3.2.8.4.1. MTT assay using both CF NP & CF-DNA NP at 37^0 C:

Cell viability of both MDAMB-231 cells and PBMC was studied by 3(4,5-dimethylthiazolyl-2)2,5- diphenyl tetrazolium bromide assay (MTT assay) according to the manufactured protocol. MDAMB-231 cells were seeded in 96-well plates followed by incubation for overnight at 37^0C in 5% CO2. After that, the cells were subjected to treatment with specific doses of CF NP and CF-DNA NP and further incubated for 24 h. Next day, the cells were washed with PBS and the media was replaced by colourless DMEM. 10 μL of MTT (5 mg/mL) was added to each well and further incubated for 6 h allowing the cleavage of MTT reagent by mitochondrial dehydrogenase of viable cells. 100 μL of DMSO was then added to each well and kept in rocking condition for 10 minutes to dissolve the formazan crystals which has purple colour. The absorbance intensity of purple colour was recorded at 570 nm on a Bio-Rad (model 550) Elisa Micro plate reader. MTT assay of PBMC cells was done by following the same protocol. Cell viability was calculated using the following formula...

$$Cell\ viability = \frac{Number\ of\ living\ cells}{Initial\ number\ of\ living\ cells} \times 100$$

3.2.8.4.2. MTT assay after magnetic hyperthermia treatment:

The cell viability rate of CF NP & CF-DNA NP treated MDAMB-231 cells was also checked after hyperthermia therapy. MDAMB-231 cells were treated with different doses of respective NPs and kept under AC magnetic field. For every dose, cells were seeded on 1 mL plate separately and incubated for overnight. Next day, cells were washed with PBS and followed by treatment with media containing CF NP & CF-DNA NP and allowed to further incubation for overnight. After that, the plates were removed from the incubator and kept inside the solenoid where AC current was applied. We have seen that our NPs were induced heat in dissolved condition under AC magnetic field. The hyperthermia therapy was applied for several times (2,5,8,11,15 minutes) on different cell plates by maintaining the plate media temperature ~ 45^0C. After hyperthermia application, the plates were removed from the solenoid and incubated at 37^0C for 24h. Next day, MTT assay was performed using the same above mentioned protocol and the absorbance intensity was recorded individually by UV-VIS spectroscopy at 570 nm.

3.2.8.4.3. In Vitro cellular uptake studies:

RITC labelled CF NP and CF-DNA NP has been used to check the internalization of these NPs inside MDAMB-231cells by using fluorescence microscope and also compare the imaging quality after DNA engineered with CF NP. 1 mg of both the RITC labelled nanoparticles were dissolved separately in 1 mL of distilled water and probe-sonicated them for proper dispersion. The MDAMB-231 cells were cultured on 12 well plates with concentration of 2×10^5cells/well and maintained for overnight. Next day, the cells further incubated for 24 h upon treatment with specific doses of both RITC labelled CF NP and CF-DNA NP. Then the images of cells were taken by fluorescence microscope (EVOS FL).

Confocal microscopy was also used to investigate the Internalization of these particles into the MDAMB-231 cells. For confocal cell imaging, the MDAMB-231 cells were seeded on cover slip and treated with same doses of RITC tagged CF NP & CF-DNA NP. Next day, the cells fixed with 3.7% formaldehyde at room temperature after washing with PBS.

3.2.8.4.4. Morphological study of cells at 37^0C and ~ 45^0C temperature:

3.2.8.4.4.1. Phase contrast microscopic image of cells:

Changes in morphology of human breast carcinoma cells upon treatment with CF NP & CF-DNA NP at two different temperature 37^0C and ~ 45^0C were initially checked using phase contrast microscope (OLYMPUS). Cells were cultured on 1 mL plates and subjected to treatment with CF NP and CF-DNA NP and allowed to incubate overnight at 37^0C. Next day, the cell morphology was observed under phase contrast microscope. To observe the changes of cell structure at ~ 45^0C, after overnight treatment, the plates were kept under AC magnetic field inside the solenoid. The cells were treated for 10 minutes at hyperthermic condition after reaching the media temperature to ~ 45^0C. After heat therapy, the plates were removed from the circuit and further incubated for 24 h. Then the cell morphology was observed under the microscope.

3.2.8.4.4.2. Scanning electron microscopic image of cell:

The morphology changes of MDAMB-231 cells upon treatment with CF-DNA NP at 37^0C & ~ 45^0C were further observed under scanning electron microscopy to re-establish the phase contrast data. The cells were grown onto cover slip in 1 mL plate and incubated at 37^0C in 5% CO_2 for overnight. Next day, the cells were treated with specific doses of CF-DNA NP and kept in incubator for 24h. To see the changes at ~ 45^0C, AC magnetic field was applied on treated cells inside the solenoid for 10 min and then re-incubated at 37^0C for overnight. Next day, the cells were fixed with 3.7% formaldehyde at room temperature for 20min after washing the cells with pre warmed PBS and followed by further washing with PBS. Then the cell samples were passed through sequential dehydration procedure of different alcohol concentrations (10%, 30%, 50%, 70%, 90% and 100%) and dried in open air. Finally, the morphological changes of MDAMB-231 cells were then observed under SEM (ZEISS, Germany)- EVO-18 special edition.

3.2.9. Experiments on cell imaging using beet root extraction fluorescent (BREF) dye:

3.2.9.1. Extraction of water soluble BREF dye:

At first, the outer layer of beet-root was pilled out and then cut into small pieces. Next, the pieces were dipped into distilled water for 1 min which causes extraction of red pigment into water. The red colour water solution was then filtered by using microfilter (0.22 μm, Millex-GP) and the filtrate was dried by rotary evaporator. After drying, the sample was washed with alcohol and acetone for several times followed by further drying. Finally, the dye pigment was stored at 4°C for months. Here, we have taken only the water soluble pigments of beet-root for cell imaging purpose.

3.2.9.2. Labelling of CF NP with BREF dye:

25 mg of CF NP was mixed with 2 mL 0.1 M $NaHCO_3$ solution and then sonicated to get a homogeneous mixture. On other side, 0.1 mg of BREF dye was dissolved in 10 mL of water from which 50 μL was dissolved in 2 mL of aqueous dimethyl sulfoxide (DMSO) followed by mixing with prior CF NP solution. Then the mixture was kept under rocking condition for 24 h in dark. Next day, the sample was collected by centrifugation and washed with water to remove excess unbound dye.

3.2.9.3. Cell imaging:

For cell imaging, we choose different cancer and normal cell lines such as breast cancer cell line MDAMB-231, lung cancer cell line A549, leukemia cells, squamous epithelium cells and PBMC. The cells (MDAMB-231, A549, leukemia, PBMC) were seeded in 6 well plates separately and incubated for 24 h. Next day the cells were treated with BREF dye and BREF functionalized CF NP followed by further incubation for 24 h. The internalization of BREF and BREF functionalized CF NP was checked by using fluorescence microscope (EVOS FL) and confocal microscope. Normal squamous epithelium cells were collected from a healthy person and directly taken in 1mL plates containing DMSO solution. The cells were treated with 20 μL of BREF dye solution and incubated for 15 min at 37°C. After incubation the cells were observed under fluorescence microscope. In case of both BREF and BREF functionalized CF NP, control solution was made by dissolving 1 mg of each sample in 1 mL PBS solution separately from which different doses has been applied on cells.

3.3. Results and Discussions:

3.3.1. Structural and morphological analysis of micelle coated CFMNPs:

Fig.3.2.(a&b) shows the SEM micrographs of two sets of CFMNPs. Here, it is observed that both the particles are spherical in shape and the average size (diameter) of the sets A and set B particles are found to be about 250 nm (Fig.3.2.(a)) and 350 nm (Fig.3.2.(b)) respectively. In case of higher concentration of oleic acid, the particle size is smaller compared to its lower concentration. At higher concentration of oleic acid, the mobility of nuclear particles becomes less therefore coagulation properties become also less which causes the smaller size particles. On other side, in case of lower concentration of oleic acid, reverse situation takes place.

Fig.3.2. SEM micrograph of CFMNPs of size (a) 250 nm and (b) 350 nm.

Fig.3.3.(a&b) show XRD pattern of the two sets of CFMNPs at room temperate. All the diffraction peaks of the pattern of both 250 nm and 350 nm CFMNPs are indexed and mapped with JCPDS data (card no. 22-1086) which indicates a good match and confirms the spinel structure of both the sample. By using the Debye Scherrer's equation ($d=0.9\lambda/(\beta\cos\theta)$), the crystallite size (d) of the samples has been calculated considering the most intense peak (311) of the XRD pattern. The crystallite size (d) of the 250 nm and 350 nm sized particles were calculated and found to be about 50 nm and 60 nm respectively.

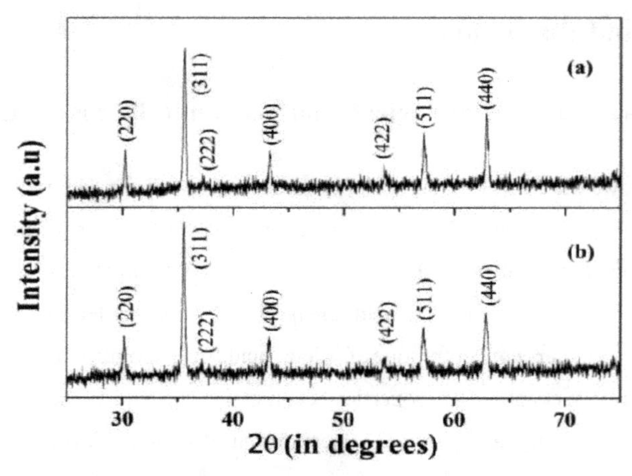

Fig.3.3. X-ray diffraction pattern of CFMNPs of size (a) 250 nm and (b) 350 nm.

3.3.2. DC and AC magnetic properties of the both CFMNPs:

To study the static magnetic property of the both CFMNPs, we have taken the DC magnetic field dependent hysteresis curve (M-H curve) of the samples at room temperature. Both the samples with particle size 250 nm and 350 nm show large hysteresis with coercivity (H_C) 1.78 KOe and 1.7 KOe respectively (Fig.3.4.(a&b)). Previous study reports that the crystallite size of a single domain $CoFe_2O_4$ particle is about 20 nm [41]. Hence, both of our synthesized CFMNPs (50 nm & 60 nm) are multi-domain nanocrystallites. Basically, in multi-domain particles magnetization dominated by motion of the domain wall which causes the decrease of coercivity with the increase of crystallite size of the particles [42]. Thus the particles of diameter 250 nm show comparatively higher coercivity than the particles of diameter 350 nm. In Fig.3.4.(a&b)), a decrease of saturation magnetization (MS) has been observed with the decrease of particle size.

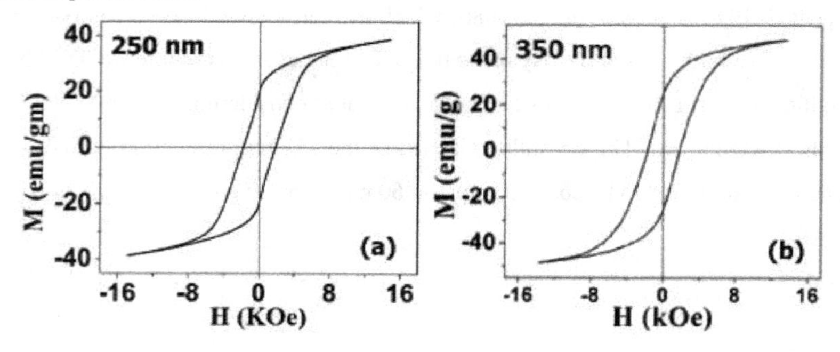

Fig.3.4. DC hysteresis loop of (a) 250 nm and (b) 350 nm CFMNPs at room temperature.

For different biological applications such as drug delivery & hyperthermia, the AC magnetic property of the CFMNPs needs to be investigated also. Under an AC magnetic field of frequency f, the hysteresis loss by the particles in one cycle of magnetization is conferred by simply integrating the hysteresis loop area which can give a quantitative measure by calculation of specific loss power (SLP) also known as SAR (specific absorption rate) value. In general, the SLP value is measured from the product of the hysteresis loop area and corresponding frequency of the applied magnetic field [43]. Thus the particles with higher value of SLP implies higher amount of heat release. To check the heat release from the CFMNPs particles, the AC hysteresis loops of the two samples were taken within a maximum magnetic field of 43 KA/m at a frequency of 50 Hz shown in Fig.3.5.(a&b). This figure revealed that the area of the hysteresis loop of the larger particles is little bit higher than that of the smaller particles. The power loss of the particles with 250 nm and 350 nm sizes was calculated from these hysteresis loops which were about 0.052 W/g and 0.061W/g respectively. Because both of the CFMNPs particles are multi-domain in character, magnetization reversal under the applied AC field accompanied by the motion of the domain wall and pinning leading to hysteresis and hence power loss [44]. In case of these particles, the SLP mainly contributed from the hysteresis loss as N′eel or Brown relaxation and frictional loss are not applicable for the ferromagnetic multi-domain solid powder system. Increase of the hysteresis loss of this particle system is related to third power law on field amplitude and linearly with the field frequency. Therefore, because of fixed field amplitude, the specific power loss varies with the square of the frequency. So, we could be obtained a much higher SAR value from these CFMNPs by increasing field and/or frequency of the AC magnetic field within the permissible limit which may be utilized to kill the cancer cells by hyperthermia therapy.

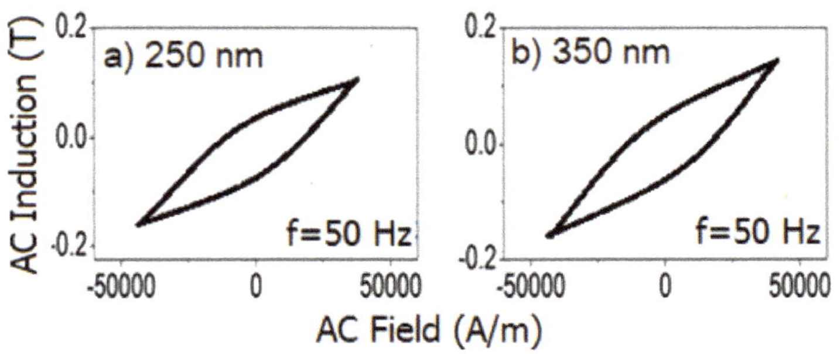

Fig.3.5. Hysteresis loop under AC magnetic field for the CFMNPs particles size of (a) 250 nm and (b) 350 nm.

3.3.3. Structural and morphological analysis of micelle frees CF NP and CF-DNA NP:

The XRD patterns of the CF NP and CF-DNA NP at room temperature represents in Fig.3.6.(a&b). By matching the all planes (220), (311), (400), (422), (511), (440) and (533) of XRD pattern with the JCPDS data (ID: 22−1086), it has been confirmed that the particles are cobalt ferrite with a cubic spinel structure. There are some changes occur in intensity of peaks in case of CF-DNA NP (Fig.3.6.(b)). This is due to incorporation of DNA with CF NP which causes changes of crystal structure of CF NP. The average crystallite size of CF NP & CF-DNA NP were calculated from Debye Scherrer eqation ($d = 0.9\lambda/\beta\cos\theta$) where, β is the full width at half maximum (FWHM) at the diffraction angle 2θ. The crystallite size of the particle was calculated using (311) peak and found to be 12.6 nm.

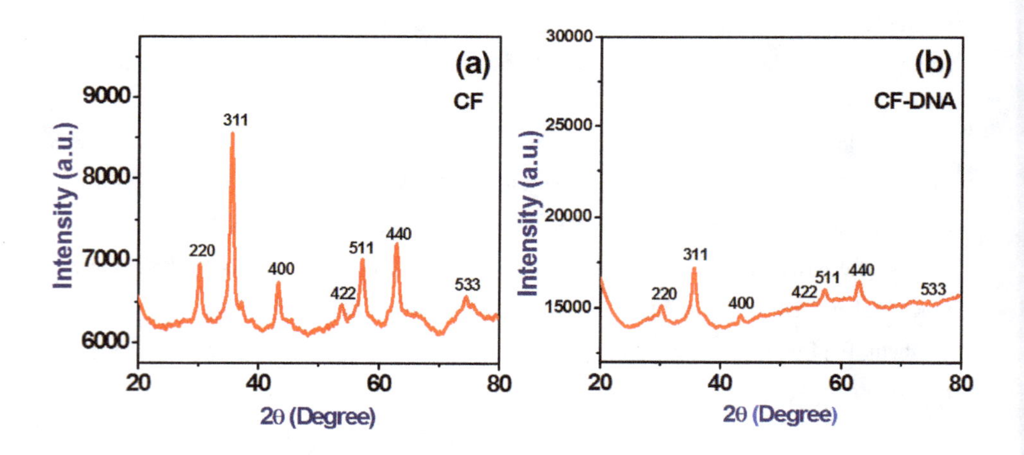

Fig.3.6. X-ray diffraction pattern of (a) CF NP and (b) CF-DNA NP

Fig.3.7.(a&b) represents the TEM images of bare CF NP and CF-DNA NP. From the TEM images, the average size of individual cobalt ferrite NP was also observed to be around 13 nm which supports the XRD data. An agglomeration has been observed in case of bare CF NP where DNA has not been used (Fig.3.7.(a)) but in case of CF-DNA NP, a chain like pattern has been observed (Fig.3.7.(b)) which is due to binding of CF NP with DNA. Basically, DNA acts as very good self-assembling agent that helps the particles arranged in different manner on the basis of DNA and precursor salt concentrations, properties of medium, synthesis procedure etc.

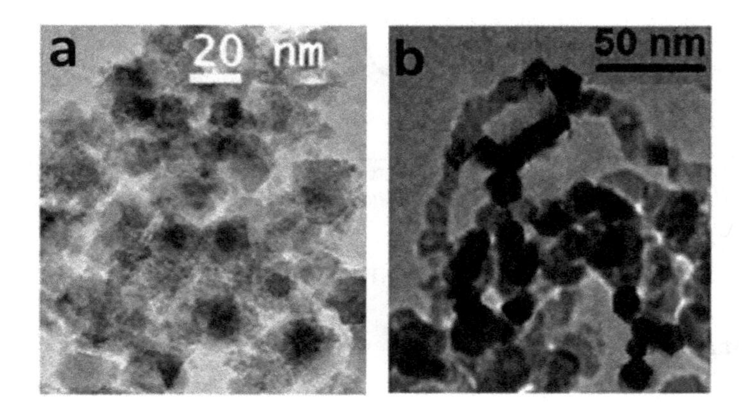

Fig.3.7. Transmission electron microscopic image of (a) CF-NP & (b) CF-DNA NP

The Fourier-transform infrared (FT-IR) spectra of bare CF NP, DNA and CF-DNA NP has been presented in Fig.3.8.(a,b,c) at the wave number range from 4000 to 400 cm^{-1}. In this figure, changes of some peak positions have been observed in CF-DNA NP compared to pure DNA. Here, some of peaks are also disappeared or some new peaks appear in CF-DNA NP with change of some pattern. These spectra are mostly matches with S. Kundu et al. study [45]. A broad peak has been observed in CF NP in range of 670 to 517 cm^{-1}, which denotes the metal oxygen in CF NP [46]. The peaks at 1715 and 1417 cm^{-1} gets from pure DNA spectrum attributed the stretching and bending vibration of C=O bond. Due to binding of DNA with CF NP, these peaks appear as one broad peak at 1632 to 1500 cm^{-1} in CF-DNA NP. In DNA spectrum, the peaks of phosphate backbone are appearing at 1240, 1080 and 825 cm^{-1}, all of which form one peak at 1170 cm^{-1} in case of CF-DNA NP. Due to binding of CF with DNA, the phosphate backbone of DNA not remains free which causes this types of changes. A broad peak at 595 to 460 cm^{-1} in CF-DNA NP attributed to binding of DNA and Co 'metal' of cobalt ferrite. The metal oxygen bond in case of CF is shifted in CF-DNA NP with change of pattern. Some new peaks are also appeared in CF-DNA NP compared to CF

NP. These results indicate the presence of DNA in nanomaterials and also support the interaction between CF and DNA.

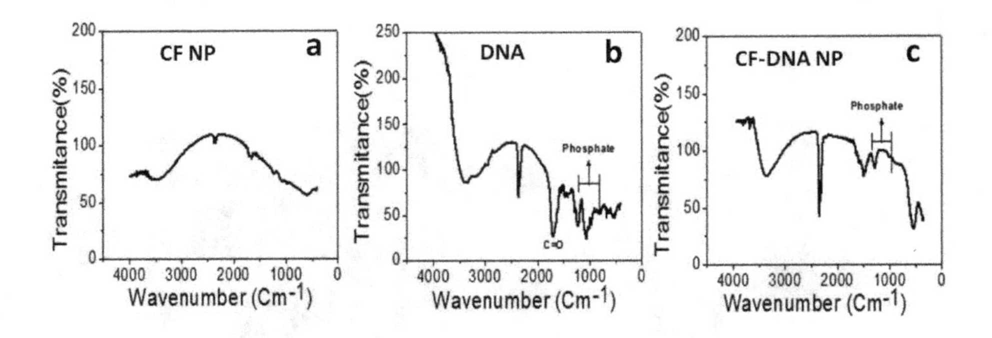

Fig.3.8. FTIR of (a) CF NP, (b) DNA, (c) CF-DNA NP

ITC data analysis:

The binding affinity between the CF NP and DNA molecule at two different temperature (37°C and 45°C) has been checked by ITC measurement and represented at Fig.3.9.(a&b). It is evident from the image that initial binding of the DNA with CF NP at 37°C temperature opens the more binding sites. At first the binding process was exothermic (Fig.3.9.(a)) and after 40 min the process is not yet totally saturated. Therefore, more binding sites were getting opened to continue this process. At higher temperature (45°C), the binding process goes through the endothermic to completely exothermic condition which is a nature favour reaction (Fig.3.9.(b)). So, this data confirms the binding of CF NP with DNA at hyperthermic temperature.

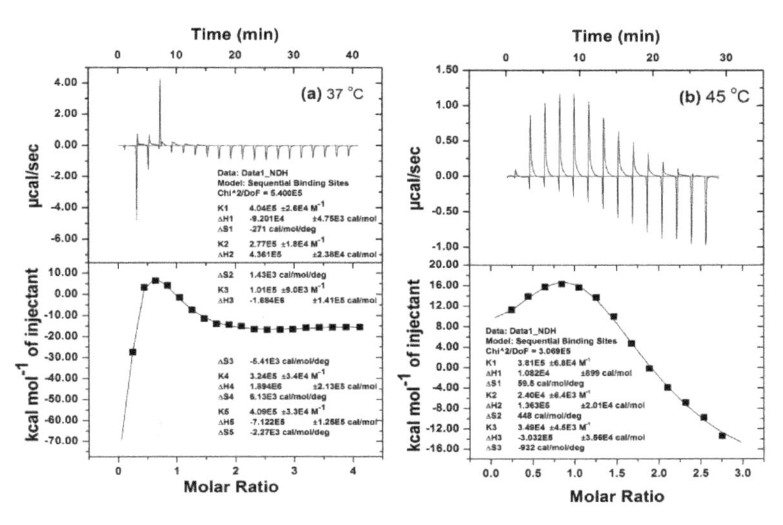

Fig.3.9. ITC represents the interaction between CF NP with DNA at (a) 37°C and (b) 45°C.

3.3.4. DC magnetic properties of CF NP & CF-DNA NP:

Fig.3.10.(a&b) represents the magnetic hysteresis loop of CF and CF- DNA NPs at 5K to 300K. Here, the coercivity (Hc) values are decreased after DNA functionalized with CF NP in all the temperatures. This happens due to ligand to metal charge transfer. Generally, DNA is diamagnetic in character and binding of diamagnetic material with ferromagnetic sample decreases the coercivity of CF-DNA NP. The coercivity of CF-DNA NP might be decreased due to weakening of anisotropy barrier which begins for thermal agitation [47]. The metal ion of CF NP is cationic in nature which make bonds with the negatively charge phosphate back bone of DNA, that can effects the anisotropy energy and causes the changing of magnetic property of CF-DNA NP [48]. The magnetization value of CF NP is compare with previous paper [49,50]. Here, the CF NP behaves like ferromagnetic which are more effective for hyperthermia application because of its good heating ability under AC magnetic field. After DNA binding, the magnetic property of CF NP getting changed, so, in this process the magnetic properties can be tuned for suitable heat generation as per required. The ferromagnetic particles have a tendency to be agglomerated up to certain level but do not hamper this particle mediated therapy. These particles were used in localized cancer treatment where dispersed particles will be directly injected in the tumour site and not be flow it all over the body when external magnetic field was applied [51].

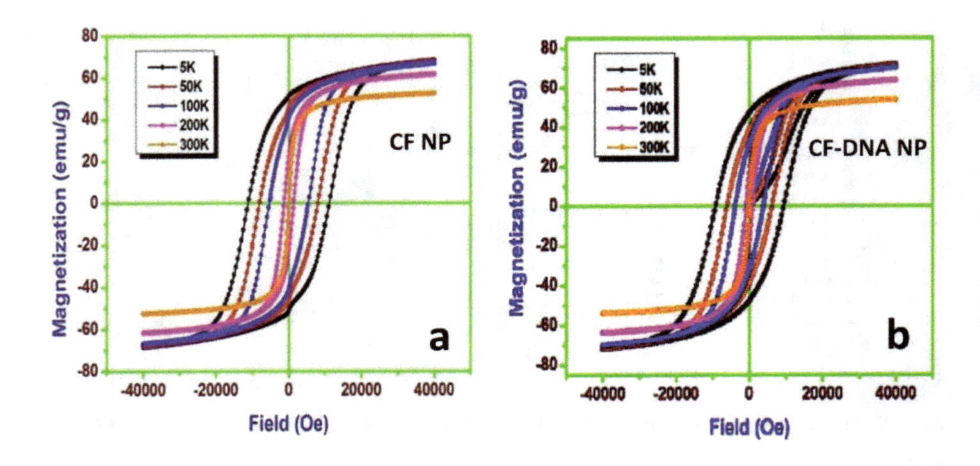

Fig.3.10. Magnetic hysteresis loop measurements at different temperatures (a) CF NP (b) CF-DNA NP.

3.3.5. AC magnetic measurements in the solenoid:

Time respected gaining and reducing temperature by the CF NP under AC magnetic field is represents in Table1 and also in Fig.3.11. The first two columns of this table represents the data related to temperature gained by the particles in dissolved condition where the last two columns illustrates the time dependent temperature losses by the particles. We observed that the CF NP gain heat quickly but they take long time to cool down. Here, we compared the data with the other 2 mL plate containing only PBS solution in which the temperature remains constant. This behaviour of CF NP is very much effective for the therapeutic treatment of cancer cells. Under AC magnetic field, the ferromagnetic materials become agitated due to frequently changes of magnetic moment which causes production of more heat for which they can be used for hyperthermia treatment.

Temp. gaining time(min.)	Temperature gained (°C) (Room temperature 24°C)	Temp. reducing time (min.) and actual time of the day	Temperature reduced (°C) (Room temperature 26°C)
1	26.00	3.23 pm	36
2	26.75	3.26 pm (3 min)	34
3	27.75	3.29 pm (6 min)	31
4	29.00	3.32 pm (9 min)	30
5	31.00	3.38 pm (15 min)	29
6	32.00	3.42 pm (19 min)	29
7	32.50	3.45 pm (24 min)	28
8	33.00	3.46 pm (25 min)	28
9	34.50	3.50 pm (29 min)	28
10	35.00	3.55 pm (34 min)	27
11	36.50	3.58 pm (37 min)	26

Table1. Reprentation of temperature gained and reduced by the CF NP with time.

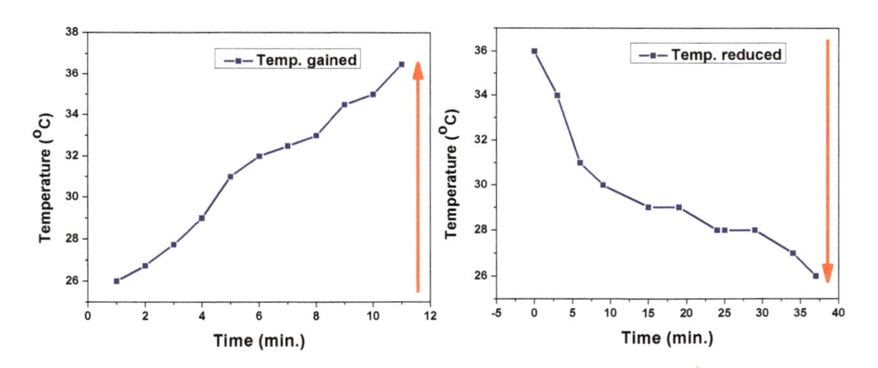

Fig.3.11. Related image of Table1 statements.

3.3.6. Studies on effects of the CF NP & CF-DNA NP on MDAMB-231 cells from different aspects:

3.3.6.1. Fluorescence imaging of MDAMB-231 cells:

Here, RITC is used as a fluorescence marker that has been tagged with both CF NP & CF-DNA NP and check their cellular uptake in human breast carcinoma cells by using fluorescent microscope (Fig.3.12.(a,b,c,d)). The fluorescent images of RITC tagged CF NP and CF-DNA NP treated cells were taken by normal fluorescent microscope (EVOS FL) represents at Fig.3.12.(a&b) as well as by confocal microscope (Olympus IX81) showing in Fig.3.12.(c&d). Here, after treatment, high fluorescence has been noticed in cytoplasm of all batches of MDAMB-231 cells and the cells which are treated with RITC tagged CF-DNA NP (Fig.3.12.(b&d)) emits more fluorescent compare to RITC tagged CF NP treated cells (Fig.3.12.(a&c)). This result proves that more CF-DNA NP has been uptake by MDAMB-231 cells than CF NP and also proves good internalization property of our particles. Therefore, DNA facilitates the interaction of CF NP with plasma membrane of cells and simplifies the cellular uptake process. As the DNA engineered CF particles shows greater intensity of the fluorescence imaging of the cells, so it can be used for cancer cell imaging with improvement of imaging quality. From previous study, it has been shown that conjugated polymers possess high absorption coefficients and high fluorescence efficiency that have a wide range of applications as fluorescent probes [52]. Hence, peoples are preferred to use of such luminescent conjugated polymers for synthesis of fluorescent nanoparticles. These nanoparticles show bright photo and electroluminescence properties after attachment with cells and also have various applications in many biological fields [53]. The conjugated backbone of such molecules gives rise to π-electron delocalization and π-π* electronic transition that allowing the formation of facilitating luminescence [54]. They are useful in the synthesis of nanoparticles for use in fluorescence imaging due to their high quantum yields and extinction coefficients in solution and also exhibit high fluorescence brightness. DNA molecules consist with a long π conjugated backbone which helps in higher population of ground state electron and therefore increase in fluorescence image intensity.

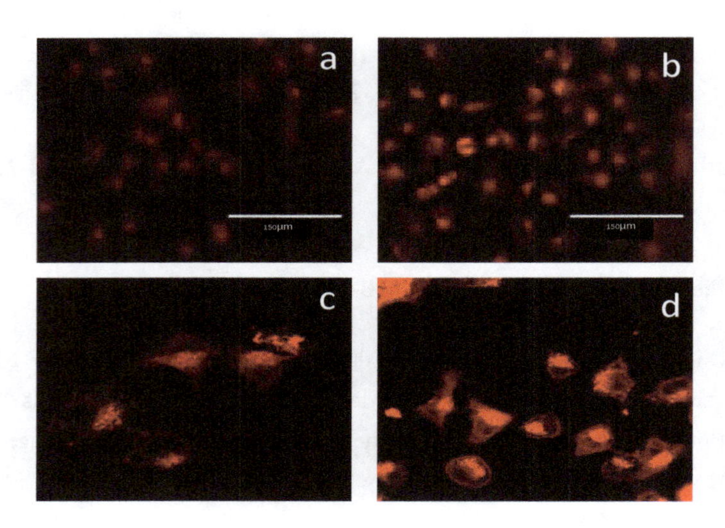

Fig.3.12. Fluorescence image of RITC tagged CF and CF-DNA NP treated MDAMB-231 cells (a&b) the normal fluorescence microscopy image of CF and CF-DNA NP respectively, (c&d) the confocal microscopy image of CF and CF-DNA NP.

We have also check the internalization of another two batches of DNA functionalized CF NP (0.2 gm and 0.8 gm DNA) by tagging them with RITC dye. A comparative study has been done to check the DNA amount at which better internalization occurs. Fig.3.13.(a&b) shows the normal fluorescent microscopic (EVOS FL) image of 0.2 and 0.8 gm DNA functionalized CF NP treated cells respectively. We observed that at same concentration of treatment, the DNA batch of 0.4 gm shows better fluorescent image compared to 0.2 gm and 0.8 gm DNA batches. Hence, the 0.4 gm DNA functionalized CF NP (CF-DNA NP) shows better cellular internalization than other batches.

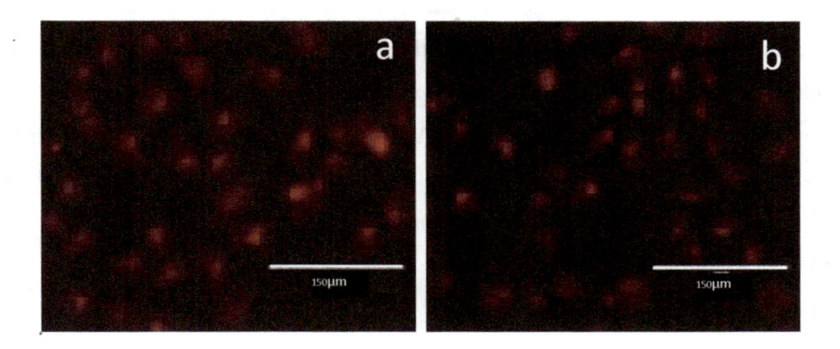

Fig.3.13. Fluorescence image of RITC tagged DNA functionalized CF NP treated MDAMB-231 cells (a) 0.2 gm DNA batch (b) 0.8 gm DNA batch

3.3.6.2. CF & CF-DNA NPs induce cell viability study at 37°C on cancer and normal cell:

MTT assay was performed to check the cytotoxicity of CF NP and CF-DNA NP on MDAMB-231 cells and PBMC represents in Fig.3.14.(a&b). CF-DNA NP has slightly higher cytotoxic effect on MDAMB-231 cells comparing with CF NP (Fig.3.14.(a)). Here we have seen that, almost 51% cell death was observed upon treatment with 35µg/ml of CF-DNA NP where CF NP causes 49% cell death at same concentration after 24h incubation. In case of CF-DNA NP, greater uptake of this NP by the cells may cause the better cytotoxic effect than CF NP. On other side, both CF & CF-DNA NP shows comparatively less toxic effect on

PBMC than MDAMB-231 cells after 24 h of treatment (Fig.3.14.(b)). Almost 32% and 35% cell death occur due to treatment with 35µg/ml of CF and CF-DNA NP respectively. From previous studies, we have seen that the NPs enter into the cell by endocytosis process and dissolution of these NPs may occur inside the lysosome of cancer cell due to its acidic environment [54]. Generally, the cancer cells pH (pH 5.5) is lower than normal cells (pH7.4) for which NPs getting dissolution in cancer cells and free ions are formed [55]. The free Fe ions produced from the NPs can goes through Fenton/ Haber- Weiss type reaction, where it reacts with hydrogen peroxide and superoxide in mitochondria and cytoplasm and produce highly reactive hydroxyl radical (.OH) and causes higher death rate of cancer cells [56]. Some other works also reports that after entering of CF NPs into the cancer cell, it become dissociated and formed free Fe^{+3} and Co^{2+} ions which induce cytotoxic effect on cancer cell [57,58]. Catelas and his group's reports [57] that the Co^{2+} ion induces apoptosis via a caspase-3 pathway and causes cell death. According to previous report, the cancer cell death is also occurs due to mitochondrial dysfunction after iron oxide NPs treatment. In cell, mitochondria plays a important role to maintain various cellular functions such as maintaining cellular homeostasis, generating ATP to provide energy etc [59,60]. In normal cells (PBMC), both CF and CF-DNA NP show less toxic effect due to its higher pH (pH7.4). Hence, both the CF & CF-DNA NPs are biocompatible for normal cells, so, it can be applied in any in vivo experiment safely.

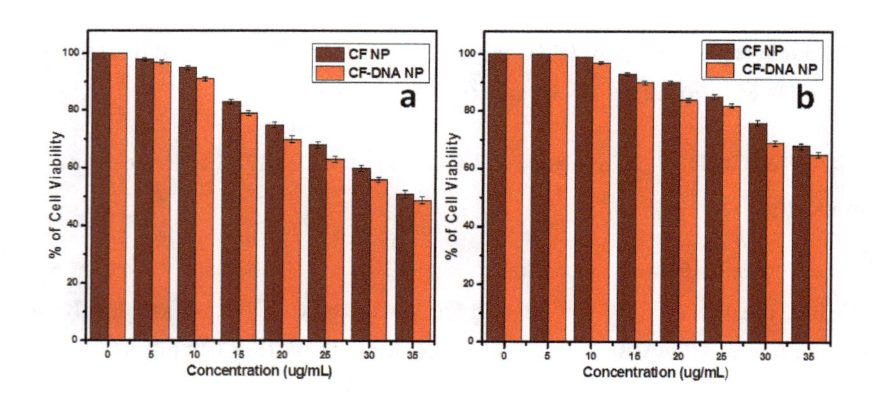

Fig.3.14. Percentage of cell viability after treatment with different doses of CF NP and CF-DNA NP in (a) MDAMB-231 and (b) PBMC cells.

3.3.6.3. CF NP & CF-DNA NP induced cell viability study on cancer cells after hyperthermia therapy (at ~45°C):

The cell viability of MDAMB-231 cells have been checked after hyperthermia treatment upon treatment with 10μg/mL of CF and CF-DNA NPs and the percentage of cells death was observed with respect of different times of heating treatment as shown in Fig.3.15. We have seen that the CF NP gained temperature very quickly in PBS solution inside the solenoid under applied current but take little long time to cool down. After hyperthermia treatment (~ 45^0C), the cell viability rate becomes decreased even at low dose of treatment (10μg/mL) compare to normal physiological temperature (37^0C) (Fig.3.14.(a)). Here, the percentage of cell death was checked with respect of 5 different times (2, 5, 8, 11, 15 min). Almost 38% and 41% cell death occurs after 15 min in case of CF NP and CF-DNA NP treated cells respectively where only 5% and 9% cell death occurs at 37^0C. Slightly higher cell death was also noticed in CF-DNA NP treated cells may be due to higher entry of this particle inside the cells. A ferromagnetic particle under AC magnetic field getting agitated and induces more heat that is the optimum condition for magnetic hyperthermia.

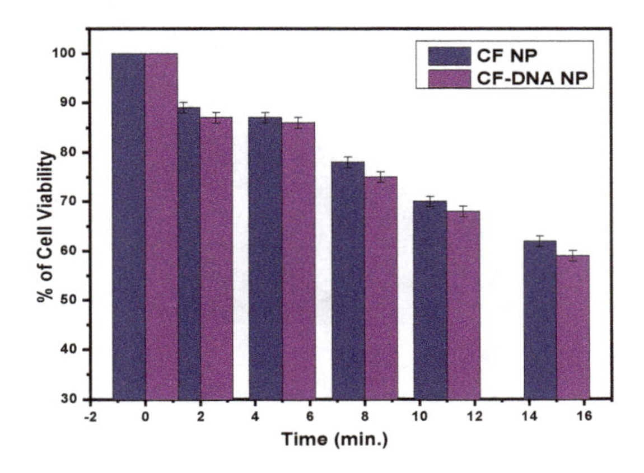

Fig.3.15. Histogram showing the percentage of cell viability of CF & CF-DNA NP treated MDAMB-231 cells after magnetic hyperthermia treatment.

3.3.6.4. Morphological changes of MDAMB-231 cells after hyperthermia treatment:

Fig.3.16. represents the phase contrast microscopic images of CF & CF-DNA NPs treated MDAMB-231 cells at normal physiological temperature (at 37^0C) and after magnetic hyperthermia treatment (at ~ 45^0C). Here, it has been observed that the cellular morphology is more disrupted after magnetic hyperthermia treatment in case of both CF & CF-DNA treated cells (Fig.3.16.(c&e)) but after incubation at 37^0C, a little changes occur in treated cells (Fig.3.16.(b&d)). The SEM images of CF-DNA treated cell at 37^0C and ~ 45^0C represents in Fig.3.17. which re-established the phase contrast data. In control set of cells, numerous protrusions have been observed on surface of the cells (Fig.3.17.(a)) by which cells make communication with other cells. But after treatment with CF-DNA NP, the protrusions started to decrease and disrupted (Fig.3.17.(b)) and after hyperthermia application the cell membrane becomes totally shrink. Hence, CF-DNA NP generates heat under AC magnetic field which destructs the normal cellular function and causes to cell death.

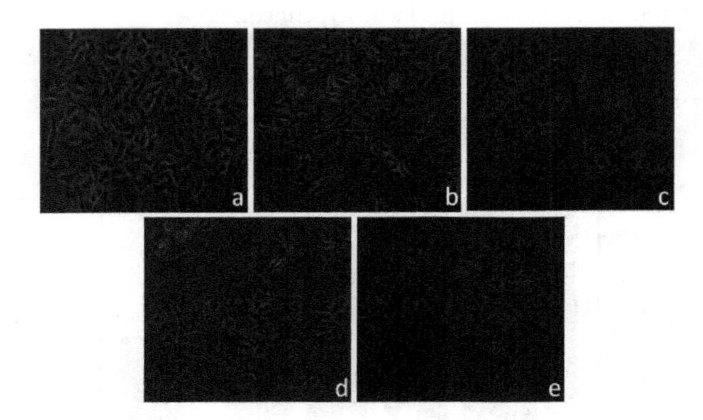

Fig.3.16. Phase contrast microscopic image of treated MDAMB-231 cells at 37^0C and ~ 45^0C temperature (a) Control (b) CF NP treated cells at 37^0C temperature (c) CF NP treated cells at ~ 45^0C temperature (d) CF-DNA NP treated cells at 37^0C temperature (e) CF-DNA NP treated cells at ~ 45^0C temperature.

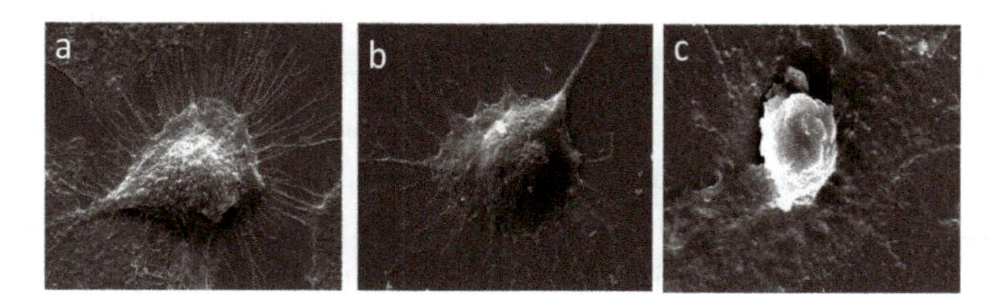

Fig.3.17. SEM image of CF-DNA NP treated cells at two different temperature (a) Control (b) CF-DNA NP treated cells at 37^{0}C temperature (c) CF-DNA NP treated cells at ~ 45^{0}C temperature.

3.3.7. Results of cell imaging experiments using BREF dye and BREF tagged CF NP:

We have checked the stability of the dye at low and high pH. Fig.3.18.(a&b) shows the smeared images of the BREF dye at pH 4.5 and pH 9 respectively. Here, we observed that at both pH the fluorescence intensity is same. So, our fluorescence dye remain stable both at higher and lower pH. Therefore, this can be used in broader pH range for many biological applications.

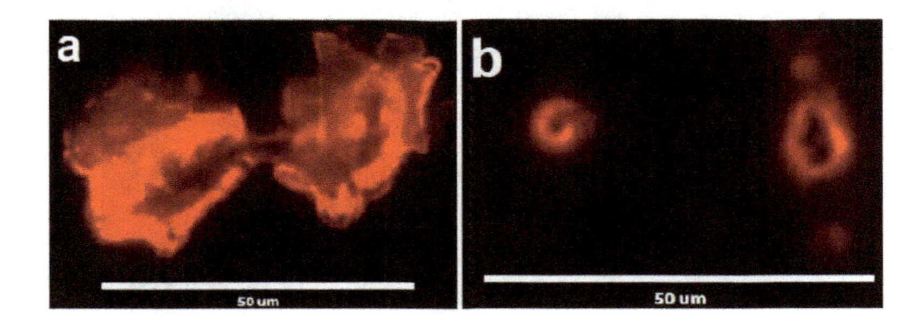

Fig.3.18. Fluorescence image of Smear BREF dye at (a) pH 4.5 and (b) pH 9

Fig.3.19.a represents the smear of dye functionalized CF NP which gives very good fluorescent image. So, the dye has successfully attached with CF NP. Fluorescence image of different treated cells represents at Fig. 3.19.(b to i). Fig.3.19.(b&c) represent the fluorescence image of BREF treated PBMC at GFP and RFP region respectively where Fig.

3.19.(d&e) represents the image of squamous epithelium cell and leukemia cell on treatment with BREF dye respectively. We incubate the squamous epithelium cell with BREF dye only for 15 min. So, from this result, it is evident that our dye takes very little time to internalize into the cells. We have also checked the internalization of BREF functionalized CF NP inside the different cells represents at Fig. 3.19.(f to i). Fig. 3.19.f is the fluorescence image of BREF functionalized CF NP treated leukemia cell where (g) & (h) are of MDAMB231 cell line at GFP and RFP region upon treatment with the BREF dye functionalized CF NP and (e) is the image of treated A549 cell. From the images, it is evident that both BREF and BREF functionalized CF NP are good fluorescent marker and can used for multiple cell imaging.

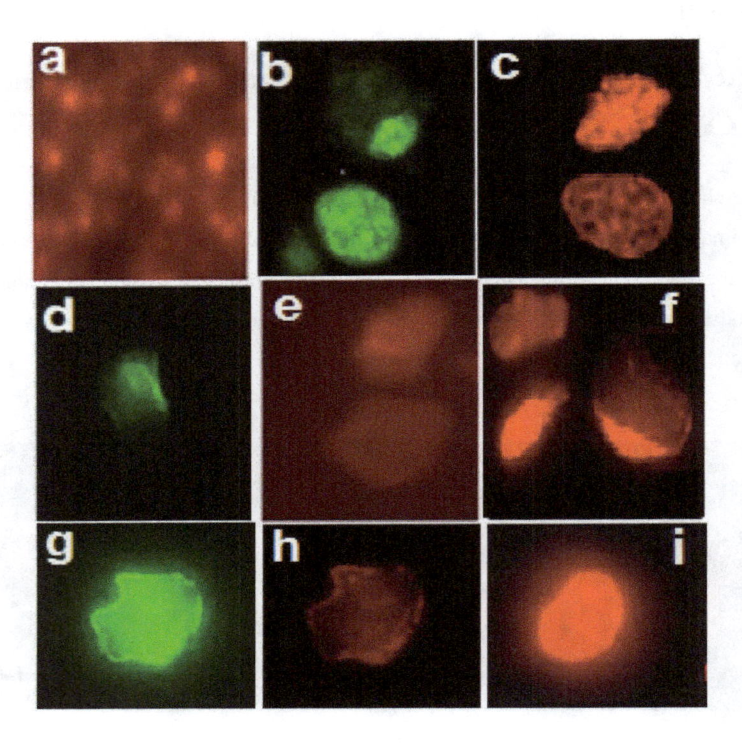

Fig.3.19. (a) Fluorescence image of smear BREF tagged CF NP. Fluorescence image of BREF treated (b) PBMC at GFP region, (c) PBMC at RFP region, (d) squamous epithelium cell, (e) leukemia cell. Fluorescence image of BREF tagged CF NP treated (f) leukemia cell, (g) MDAMB-231 at GFP region, (h) MDAMB-231 at RFP region, (i) A549 cell.

We have also take confocal microscopic image of MDAMB-231 cell line after treating them with free BREF and BREF functionalized CF NP at GFP and RFP regions (Fig.3.20.). Fig. 3.20.(a&b) are the dye treated cells and Fig.3.20.(c&d) are dye tagged CF NP treated cells. All are show good fluorescent images at both GFP and RFP regions. Hence, we can say that the BREF dye extracted by our method is very good marker for cell imaging. It is known that majority of the pigment present in red beet root is betanin (75 to 95%), is a water soluble dye which is mainly responsible for fluorescence property. Another water insoluble pigment betaxanthine is also present in red beet root but in very low quantity [61]. Here, we concentrate on water soluble pigment only because for biological cell imaging purposes, water solubility of dye is more preferable.

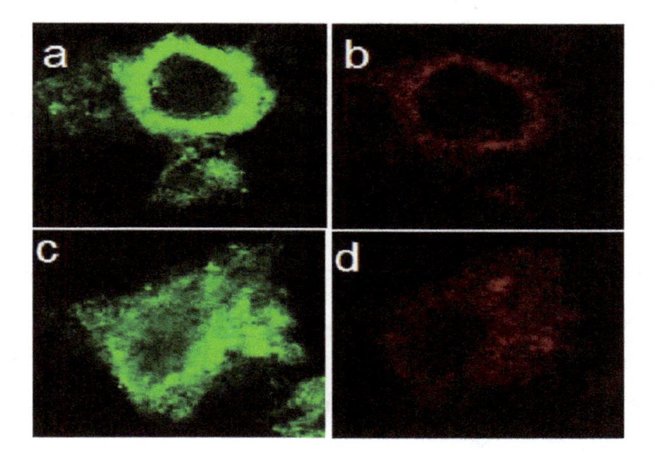

Fig.3.20. Confocal microscopic image of MDAMB-231 cell lines after treating them with (a) & (b) BREF dye at GFP & RFP region respectively, and (c) & (d) BREF tagged CF NP at GFP & RFP respectively.

3.4. Conclusion:

In summary, we have successfully synthesized the spherical Cobalt ferrite nanoparticles (CFMNPs) of size about 250 nm and 350 nm by chemical co-precipitation method using oleic acid as micelles. Structural characterization of these particles confirms the particles are pure phase cobalt ferrite. The crystallite size (d) of 250 nm and 350 nm size particles are to be about 50 nm and 60 nm respectively. Hence, the particles are multi-domain in nature (the d of a single domain particle is about 20 nm). Here, we observed that the particle size become changed upon the concentration of oleic acid and at higher concentration of oleic acid, the particle's size is smaller and vice versa. Both the samples of particle size 250 nm and 350 nm exhibit good DC magnetic property at room temperature and shows large hysteresis with coercivity (HC) 1.78 KOe and 1.7 KOe respectively. The particles with size of 250 nm show comparatively higher coercivity than the particles size of 350 nm. Both the particles also show good AC magnetic property and specific loos power. This study indicates about the crucial importance of these particles in hyperthermia therapy.

Then, we have synthesized the single domain cobalt ferrite nanoparticle and DNA engineered cobalt ferrite nanoparticles by simple chemical co-precipitation method and it was confirmed that cobalt ferrite is successfully engineered on DNA scaffold. Structural characterization of these particles proves the existence of cobalt ferrite is in its pure phase. These materials act as a very good fluorescent agent for cell imaging after tagging them with fluorescent marker. The fluorescence intensity has been increased inside the cell in case of CF-DNA NP compare to CF NP. Therefore, these particles can be used for biomedical applications. More CF-DNA NP can enter into the cells than CF NP and shows slight higher cytotoxic effect on cells. These particles are biocompatible for PBMC. The magnetic property of CF NP has been changed after DNA engineering which helps to tune the magnetic property for suitable heat generation as per required. Here, we observed that the CF NP produced heat under AC magnetic field and after treatment, more MDAMB-231 cell death occurs due to hyperthermia application. The cell morphology was also changed upon treatment and cells become shrink after hyperthermia therapy. So, our particles has been used for magnetic hyperthermia successfully and also used for cell imaging.

Next, we have extracted the dye from beet root by very easy process which has a very good fluorescent property. The dye has been successfully tagged with CF NP and both the free dye and dye functionalized CF NP gives very good fluorescence image of cancer and normal cells. This dye has stability in both low and high pH, so it can be used for different biological purposes. This dye can enter into the cells within 15 min. Hence, it can be useful for instant cell imaging. Because this dye comes directly from natural organic substance (beet root), it is well known nontoxic and anti-oxidant in nature. So, our dye is very cost effective compared to other cell imaging dye and it has been used for multiple cell imaging.

Bibliography:

[1] Lu, Y., He, B., Shen, J., Li, J., Yang, W., Yin, M., 2015. Multifunctional magnetic and fluorescent core-shell nanoparticles for bioimaging. Nanoscale, 7, 1606–1609.

[2] Tietze, R., Zaloga, J., Unterweger, H., Lyer, S., Friedrich, R.P., Janko, C., Pöttler, M., Dürr, S., Alexiou, C., 2015. Magnetic nanoparticle-based drug delivery for cancer therapy Biochem. Biophys. Res. Commun., 468, 463–470.

[3] Mody, V.V., Cox, A., Shah, S., Singh, A., Bevins, W., Parihar, H., 2014. Magnetic nanoparticle drug delivery systems for targeting tumor. Appl. Nanosci., 4, 385–392.

[4] Goswami, M.M., Dey, C., Bandyopadhyay, A., Sarkar, D., Ahir, M., 2016. Micelles driven magnetite (Fe3O4) hollow spheres and a study on AC magnetic properties for hyperthermia application. J. Magn. Magn. Mater., 417, 376–381.

[5] Dey, C., Baishya, K., Ghosh, A., Goswami, M.M., Ghosh, A., Mandal, K., 2017. Improvement of drug delivery by hyperthermia treatment using magnetic cubic cobalt ferrite nanoparticles. J. Magn. Magn. Mater., 427, 168–174.

[6] Pankhurst, Q.A., Connolly, J., Jones, S.K., Dobson, J., 2003. Applications of magnetic nanoparticles in biomedicine. J. Phys. D. Appl. Phys., 36, R167.

[7] Chen, J., Guo, Z., Tian, H., Chen, X., 2016. Production and clinical development of nanoparticles for gene delivery. Mol. Ther. - Methods Clin. Dev., 3, 16023.

[8] McCarthy, J.R., Weissleder, R., 2008. Multifunctional magnetic nanoparticles for targeted imaging and therapy. Adv. Drug Deliv. Rev., 60, 1241–1251.

[9] Yang, H.W., Hua, M.Y., Liu, H.L., Huang, C.Y., Wei, K.C., 2012. Potential of magnetic nanoparticles for targeted drug delivery. Nanotechnol. Sci. Appl., 5, 73–86.

[10] Bao, F., Yao, J.L., Gu, R.A., 2009. Synthesis of Magnetic Fe2O3/Au core/shell nanoparticles for bioseparation and immunoassay based on surface-enhanced Raman spectroscopy. Langmuir, 25, 10782–10787.

[11] Meddahi-Pelle, A., Legrand, A., Marcellan, A., Louedec, L., Letourneur, D., Leibler, L., 2014. Organ Repair, Hemostasis, and In Vivo Bonding of Medical Devices by Aqueous Solutions of Nanoparticles. Angew. Chem Int. Ed., 53, 6369–6373.

[12] http://www.medicalnewstoday.com/articles/158401.php, http://www.cancer.org/treatment/treatmentsandsideeffects/index?sitearea=

[13] Andra, W., Nowak, H., 2007. Magnetism in Medicine: A Handbook, Second Edition. Wiley-VCH Verlag GmbH & Co. KGaA.

[14] Kobayashi, T., 2011. Cancer hyperthermia using magnetic nanoparticles. J. Biotechnology, 6, 1342-1347.

[15] Jordan, A., Scholz, R., Wust, P., Fahling, H., Felix, R., 1999. Magnetic fluid hyperthermia (MFH): Cancer treatment with AC magnetic field induced excitation of biocompatible superparamagnetic nanoparticles. J. Magnetism and Magnetic Materials, 201,413-419.

[16] Maenosono, S., Saita, S., 2006. Theoretical assessment of FePt nanoparticles as heating elements for magnetic hyperthermia. IEEE Trans. On Magnetics, 42 (6), 1638-1642.

[17] Lagendijk, J.J.W., 2000. Hyperthermia treatment planning. Phys. Med. Biol., 45, R61.

[18] Jordan, A., Scholz ,R., Maier-Hauff, K., Johannsen, M., Wust, P., Nadobny, J., Schirra, H., Schmidt, H., Deger, S., Loening, S., Lanksch, W., Felix, R., 2001. Presentation of a new magnetic field therapy system for the treatment of human solid tumors with magnetic fluid hyperthermia. J Magnetism Magn Mater, 225, 118–126.

[19] Maier-Hauff, K., Rothe, R., Scholz, R., Gneveckow, U., Wust, P., Thiesen, B., Feussner, A., Deimling, A.V., Waldoefner, N., Felix, R., Jordan, A., 2007. Intracranial thermotherapy using magnetic nanoparticles combined with external beam radiotherapy: results of a feasibility study on patients with glioblastoma multiforme. J. Neurooncol., 81 (1), 53-60.

[20] Huang, C.F., Lin, X.Z., Lo, W.H., 2010. Design and construction of a hyperthermia system with improved interaction of magnetic induction-heating. Annual International Conference of the IEEE Engineering in Med Bio Soc., 2010, 3229–3232.

[21] Hugounenq, P., Levy, M., Alloyeau, D., Dubois, E., et al., 2012. Iron Oxide Monocrystalline Nanoflowers for Highly Efficient Magnetic Hyperthermia. J.Phys. Chem. C, 116 (29), 15702-15712.

[22] Ruiz-Hernandez, E., Baeza, A., Vallet-Regi, M., 2011. Smart drug delivery through DNA/magnetic nanoparticle gates. ACS Nano, 5 (2), 1259–1266.

[23] Sonvico, F., Mornet, S., Vasseur, S., Dubernet, C., Jaillard, D., Degrouard, J., Hoebeke, J., Duguet, E., Colombo, P., Couvreur, P., 2005. Folate-conjugated iron oxide nanoparticles for solid tumor targeting as potential specific magnetic hyperthermia mediators: synthesis, physicochemical characterization, and in vitro experiments. Bioconjug. Chem., 16, 1181–1188.

[24] Yang, W.W., Pierstorff, E., 2012. Reservoir-Based Polymer Drug Delivery Systems. J. Lab. Autom., 17, 50–58.

[25] Campelj, S., Makovec, D., Drofenik, M., 2008. Preparation and properties of water-based magnetic fluids. J. Phys. Condens. Matter, 20, 204101.

[26] Nguyen, D.T., Kim, K.S., 2016. Controlled synthesis of monodisperse magnetite nanoparticles for hyperthermia-based treatments. J. Powder Technology, 301, 1112-1118.

[27] Jordan, A., Scholz, R., Wust, P., Schirra, H., Schiestel, T., Schmidt, H., Felix, R., 1999. Endocytosis of dextran and silan-coated magnetite nanoparticles and the effect of intracellular hyperthermia on human mammary carcinoma cells in vitro. J. Magnetism and Magnetic Materials, 194, 185-196.

[28] Zhang, L.Y., Gu, H.C., Wang, X.M., 2007. Magnetite ferrofluid with high specific absorption rate for application in hyperthermia. J. Magnetism and Magnetic Materials, 311, 228-233.

[29] Makridis, A., Chatzitheodorou, I., Topouridou, K., Yavropoulou, M.P., Angelakeris, M., Dendrinou-Samara, C. A., 2016. Facile microwave synthetic route for ferrite nanoparticles with direct impact in magnetic particle hyperthermia. J. Materials Science and Engineering: C, 63, 663-670.

[30] Çelik, O., Can, M.M., Firat, T., 2014. Size dependent heating ability of CoFe2O4 nanoparticles in AC magnetic field for magnetic nanofluid hyperthermia. J. Nanoparticle Research, 16, 1-7.

[31] Salunkhe, A.B., Khot, V.M., Ruso, J.M., Patil, S.I., 2016. Water dispersible superparamagnetic Cobalt iron oxide nanoparticles for magnetic fluid hyperthermia. J. Magnetism and Magnetic Materials, 419, 533-542.

[32] Das, H., Sakamoto, N., Aono, H., Shinozaki, K., Suzuki, H., Wakiya, N., 2015. Investigations of superparamagnetism in magnesium ferrite nano-sphere synthesized by ultrasonic spray pyrolysis technique for hyperthermia application. J. Magnetism and Magnetic Materials, 392, 91-100.

[33] Khot, V.M., Salunkhe, A.B., Thorat, N.D., Ningthoujam, R.S., Pawar, S.H., 2013. Induction heating studies of dextran coated MgFe2O4 nanoparticles for magnetic hyperthermia. J. Dalton Transactions, 42, 1249-1258.

[34] Iftikhar, A., Islam, M.U., Awan, M.S., Ahmad, M., Naseem, S., Iqbal, M.A., 2014. Synthesis of super paramagnetic particles of Mn1−xMgxFe2O4 ferrites for hyperthermia applications. J. Alloys and Compounds, 601, 116-119.

[35] Hanini, A., Lartigue, L., Gavard, J., Kacem, K., Wilhelm, C., Gazeau, F., Chau, F., Ammar, S., 2016. Zinc substituted ferrite nanoparticles with Zn0.9Fe2.1O4 formula used as

heating agents for in vitro hyperthermia assay on glioma cells. J. Magnetism and Magnetic Materials, 416, 315-320.

[36] Giri, J., Pradhan, P., Sriharsha, T., Bahadur, D., 2005. Preparation and investigation of potentiality of different soft ferrites for hyperthermia applications. J. Applied Physics., 97, 10Q916.

[37] Jing, Y., Sohn, H., Kline, T., Victora, R.H., Wang, J.P., 2009. Experimental and theoretical investigation of cubic FeCo nanoparticles for magnetic hyperthermia. J. Applied Physics., 105, 07B305.

[38] Kuznetsov, A.A., Leontiev, V.G., Brukvin, V.A., Vorozhtsov, G.N., Kogan, B.Y., Shlyakhtin, O.A., Yunin, A.M., Tsybin, O.I., Kuznetsov, O.A., 2007. Local radiofrequency induced hyperthermia using CuNi nanoparticles with therapeutically suitable Curie temperature. J. Magnetism and Magnetic Materials, 311, 197-203.

[39] Sun, S., Zeng, H., Robinson, D.B., Raoux, S., et al., 2004. Monodisperse MFe_2O_4 (M = Fe, Co, Mn) Nanoparticles. J. Am.Chem. Soc., 126 (1), 273-279.

[40] Joshi, H.M., Lin, Y.P., Aslam, M., Prasad, P.V., et al., 2009. Effects of shape and size of cobalt ferrite nanostructures on their MRI contrast and thermal activation. J. Phys.Chem. C, 113 (41), 17761-17767.

[41] Pal, D., Mandal, M., Chaudhuri, A., Das, B., et al., 2010. Micelles induced high coercivity in single domain cobalt-ferrite nanoparticles. J. Appl.Phys, 2010; 108, 124317.

[42] Cullity, B.D.,1972. Introduction to Magnetic Material, 2nd Edition. London: Addison-Wesley.

[43] Goswami, M.M., 2016. Synthesis of Micelles Guided Magnetite (Fe_3O_4) Hollow Spheres and their application for AC Magnetic Field Responsive Drug Release. Sci. Reports, 6, 35721.

[44] Rudolf, H., Silvio, D., Robert, M., Matthias, Z., 2006. Magnetic particle hyperthermia: nanoparticle magnetism and materials development for cancer therapy. J. Phys.:Condens. Matt., 18, S2919-S2934.

[45] Nithiyanantham, U., Ramadoss, A.K., Ede, S. R., Kundu, S., 2014. DNA mediated wire-like clusters of self-assembled TiO 2 nanomaterials: supercapacitor and dye sensitized solar cell applications. Nanoscale, 6 (14) 8010–8023.

[46] Ristic, M., Krehula, S., Reissner, M., Jean, M., Hannoyer, B., Music, S., 2016. Journal of Molecular Structure S0022-2860 (16), 31005.

[47] Dey, C., Baishya, K., Ghosh, A., Goswami, M.M., Ghosh, A., Mandal, K., 2017. Improvement of drug delivery by hyperthermia treatment using magnetic cubic cobalt ferrite nanoparticles. J. Magn. Magn. Mater, 427, 168–174.

[48] Sarkar, D., Mandal, M., 2012. Static and dynamic magnetic characterization of DNA templated chain like magnetite nanoparticles. J. Phys. Chem. C, 116 (5), 3227–3234.

[49] Maaz, K., Mumtaz, A., Hasanain, S.K., Ceylan, A., 2007. Synthesis and magnetic properties of cobalt ferrite ($CoFe_2O_4$) nanoparticles prepared by wet chemical route. J. Magn. Magn. Mater, 308, 289–295.

[50] Chen, R., Christiansen, M.G., Anikeeva, P., 2013. Maximizing Hysteretic Losses in Magnetic Ferrite Nanoparticles via Model-Driven Synthesis and Materials Optimization. ACS Nano, 7 (10), 8990–9000.

[51] Javidi, M., Heydari, M., Karimi, A., Haghpanahi, M., Navidbakhsh, M., Razmkon, A., 2014. Evaluation of the effects of injection velocity and different gel concentrations on nanoparticles in hyperthermia therapy. J. Biomed. Phys. Eng., 4, 151–162.

[52] Wu, C., Szymanski, C., Cain, Z., McNeill, J., 2007. Conjugated polymer dots for multiphoton fluorescence imaging. J Am Chem Soc., 129(43), 12904–12905.

[53] (a) Schwartz, B.J., 2003. How Chain Conformation and Film Morphology Influence Energy Transfer and Interchain Interactions. Annu Rev Phys Chem, 54, 141–172. (b) Wu, C., Szymanski, C., Cain, Z., McNeill, J., 2007. Conjugated polymer dots for multiphoton fluorescence imaging. J Am Chem Soc., 129, 12904–12905. (c) Wu, C., Bull, B., Szymanski, C., Christensen, K., McNeill, J., 2008. Multicolor Conjugated Polymer Dots for Biological Fluorescence Imaging. ACS Nano., 2, 2415–2423.

[54] Su, J., Chen, F., Cryns, V.L., Messersmith, P.B., 2011. Catechol Polymers for pH-Responsive, Targeted Drug Delivery to Cancer Cells. J Am Chem Soc., 133 (31), 11850–11853.

[55] De, D., Goswami, M.M., 2016. Shape induced acid responsive heat triggered highly facilitated drug release by cube shaped magnetite nanoparticles. Biomicrofluidics, 10, 064112.

[56] Wu, H., Yin, J.J., Wamer, W.G., Zeng, M., Lo, Y.M., 2001. Reactive oxygen species-related activities of nano-iron metal and nano-iron oxides. Journal of food and drug analysis, 22, 86 -94.

[57] Catelas, I., Petit, A., Zukor, D.J., Huk, O.L., 2001. Cytotoxic and apoptotic effects of cobalt and chromium ions on J774 macrophages – Implication of caspase-3 in the apoptotic pathway. J. Mater. Sci. Mater. Med., 12, 949-953.

[58] Kwon, Y-M., Xia, Z., Glyn-Jones, S., Beard, D., Gill, H.S., Murray, D.W., 2009. Dose-dependent cytotoxicity of clinically relevant cobalt nanoparticles and ions on macrophages in vitro. Biomed Mater, 4 (2), 25018.

[59] Soriano, M.E., Scorrano, L., 2011. Traveling Bax and forth from mitochondria to control apoptosis. Cell Elsevier Inc., 145 (1), 15-7.

[60] Green, D.R., Galluzzi, L., Kroemer, G., 2011. Mitochondria and the autophagy-inflammation-cell death axis in organismal aging. Science, 333 (6046), 1109–1112.

[61] Sing, B., Hathan, B.S., 2014. Chemical composition, functional properties and processing of beetroot —a review. Int. J. Sci. Eng. Res., 5, 679–784.

Chapter 4

Use of magnetic nanoaprticles for drug delivery

In this chapter we have discussed about the synthesis of magnetic nanoparticles (MNPs) and their surface modification. Here we have focused on drug delivery by using magnetic nanoparticls in different cancer cells. For this work, characterization of these MNPs has been done in proper way and the efficiency of drug release and efficacy of drug activity have been studied in detail at cellular level. Here, the drug release from MNPs was also checked under different stimuli such as heat and pH.

4.1. Induction:

Nanomaterials give an important attention on researcher's mind from the last few decades due to their unique characteristics over their usual bulk character [1-6]. The magnetic nanoparticles (MNPs) offer various applications in biological field such as in drug delivery [7], hyperthermia [8], bio-separation [9], magnetic resonance imaging [10], biosensor [11] etc [12]. It is known that, cancer is one of the harmful diseases in the world and there are many types of cancer treatments used to treat cancers such as surgery, radiation therapy, chemotherapy etc. but those methods have several side effects. Besides of these treatments, nanoparticle based drug delivery systems (DDS) gives an attention in research area due to its more efficiency for cancer treatment [13,14]. For biological use, the MNPs must have some important properties such as tunable size, large surface area, proper functionalization, magnetic property, good dispersion and high stability in liquid medium, non-toxicity etc.

Many anticancer drugs have been developed showing limitation in their clinical use because of their low selectivity, non specificity, poor aqueous solubility and low bioavailability. The researchers have overcome these problems by synthesizing the nanoparticles and functionalized them in such a way that helps a sustained and a controlled release of its cargo at cancer sites. MNPs based anticancer drug delivery system were used for their property of enhanced permeation retention effect into which the MNPs are accumulated and retained in the cancerous extracellular matrix for longer time [15-17]. Due to the poor cellular internalization of drugs, many times chemotherapy becomes non efficient which limits the drug activity and dosages of anticancer drugs administered for treatment become below its therapeutic level [18,19]. Hence, to overcome this problems, controlled release of drugs at cancer sites have been tried. In addition with MNPs, recently different stimuli also play important roles in possible drug delivery. The common stimuli are pH,[20] temperature,[21,22] magnetic property,[8] enzyme [23] etc. which perform potent role in drug delivery. Among them pH responsive drug released is very important due to difference of pH between different tissues inside the body. The pH of our normal body tissues is slightly alkaline (pH7.4), whereas tumor tissue maintains pH around acidic environment (pH5.5) [24]. Various pH responsive MNPs based targeted deliveries have been developed for better drug delivery [25-29]. Recently researchers are trying to use anti angiogenesis drugs for cancer treatment. Generally cancer becomes more fatal when it starts to metastasis and this is usually depending on angiogenesis. So, use of anti-angiogenesis drug for cancer treatments receipt an interest for clinical therapy [30,31]. Besides the DDS, the MNPs can also be able to serve as drug carrier and drug delivery agent by magnetically triggered way due to magnetic nature of these particles [32-37]. In that case, the release of drugs from the MNPs was controlled by magnetically induced heat treatment pathway. In previous chapter, we have already shown that magnetic nanoparticles release heat under alternating current (AC) magnetic field and this heat was successfully able to destroy the cancer cells. On the other hand, we have proposed how alternating current (AC) magnetic field will trigger the drug release from magnetic particles which can be controlled from outside the body and also shown the combined effect of pH and heat on drug release. The magnetic nanoparticles can convert the magnetic energy into thermal energy which may increase potentiality to use of these MNPs for therapeutic purpose [38-40]. Under an external alternating current (AC) magnetic field, the magnetic particles are heated up and this response can be obtained from the specific loss power (SLP) measurement [41]. Another advantage of use of the MNPs is that, it can be monitored in a non-invasive way from outside the body by directing and controlling the magnetic field required. From previous studies, it has been noticed that the shape of the particles plays an important role for use of these particles as drug delivery agents [42,43]. Some previous studies reports that the particles other than spherical shape are inhibited to be recognized by the immune system. In addition, the proper particle shape can enhance the cellular interaction and the circulation time of the drug carrier [44]. Hence, particles other than spherical shape has importance for drug delivery and drug release because these particles will be body compatible and also have some other properties such as magnetic, heating, flow rate in blood, stability, etc., which will increase the scope to modify the drug delivery and release in the body.

In our study, we have synthesized the cube shaped ferro-magnetic magnetite (Fe_3O_4) nanoparticles by very simple method. Here, doxorubicin (DOX) drug has been used as an anti-cancer drug. We have reported that this particle will serve as a very good nano-carrier for drug delivery system under different stimuli such as pH and temperature. Here, we investigate the combined effect of these two stimuli i.e. heat and pH simultaneously on efficiency of anti cancer drug release. We have studied the AC magnetic properties of Fe_3O_4 nanoparticles and also checked the heat generation under AC frequency.

Fruthermore, we have synthesized the spherical shaped two sizes of cobalt ferrite nanoparticles (CFMNPs) whose details characterization and AC; DC magnetic properties are discussed in chapter 3. Here, we have reported the drug release studies from these CFMNPs under AC magnetic field.

Next, we have studied the delivery of dopamine (DA) as an anti-cancer drug at A549 cell lines by using cobalt ferrite (CF) nanoparticles as a drug carrier. DA is highly water soluble which causes easy release of DA into the solution after binding with CF. Therefore; PEG_{4000} has been used to stabilize the binding of CF with DA and produced CF-DA-PEG. Generally, the nanocarrier was entrapped by reticuloendothelial system (RES) inside the body which decreased the efficiency of this nanocarrier to reach the target cells [45] but functionalization of nanocarrier with PEG stops the RES opsonization and generates a hydrophilic barrier that enhances stabilization when circulating into blood [46]. In this study, the CF-DA-PEG has been used as a carrier for DA delivery into A549 cells. Here, we have checked the effect of CF-DA-PEG on ROS generation and apoptosis on A549 cells. We also observed the changes of mitochondrial membrane potential after treatment with CF-DA-PEG. The cell viability was studied by MTT assay. The anti-migratory effect of this particle on A549 has been investigated by different assay where the cell morphology was observed by using SEM. After treatment, different protein expression of cells has been checked by performing western blot.

4.2. Experimental details:

4.2.1. Materials:

The following reagents required for synthesis of nanoparticles as mentioned below:

Ferric chloride hexahydrate ($FeCl_3.6H_2O$; purity 98%), Cobalt chloride hexahydrate ($CoCl_2.6H_2O$; purity 97%), Iron acetyl acetonate, cobalt acetate, Rodamin B- isothiocyanate (RITC), ethylene glycol, urea, Tx-100, oleylamine, phenyl ether, oleic acid and doxorubicin hydrochloride were procured from Sigma Aldrich. Potassium hydroxide (KOH), phosphate buffer solution (PBS), Dimethyl sulfoxide (DMSO), ethanols were purchased from Merck, Germany. The chemicals for cell culture such as culture media Dulbecco's Modified Eagle's Medium (DMEM), Roswell Park Memorial Institute 1640 (RPMI-1640), non essential amino acids, amphotericin B, fetal bovine serum (FBS), penicillin, streptomycin, gentamycin, L-glutamine etc. were purchased from HIMEDIA (Mumbai, India). Antibodies of Caspase3, Caspase9, cytochrome c, p^{53}, BaX, Bcl2 were obtained from Santa Cruz Biotechnology (Santa Cruz, CA). DAPI and Hoechst 123 were bought from Invitrogen (Carlsbad, CA, USA).

4.2.2. Synthesis of magnetite (Fe_3O_4) nanoparticle:

The synthesis procedure of Fe_3O_4 MNPs is as follows: 4.05 gm of $FeCl_3.6H_2O$ was dissolved in 70 mL ethylene glycol. Then 10 gm of Urea was added to this parent solution and stirred vigorously to get a homogeneous mixture. After that, 2 mL of Tx-100 was added to the mixture and again stirred for 6 h along with heat. After about 1 h of stirring, the solution colour became dark yellowish when 2 mL of oleylamine was added to this solution. While 6 h of stirring and heating was completed, the solution was allowed to cool at room temperature and kept for overnight after doubling the solution volume with ethanol. Next day, the solution was washed for several times with de-ionized water, ethanol and acetone and collected by centrifugation. The samples were dried at air and ground well in a mortar. Before characterizations, the sample was annealed at 300^0C for 1 h under nitrogen atmosphere.

4.2.3. Synthesis of micelle cobalt ferrite nanoparticle (CFMNPs):

In previous chapter (chapter 3), we have already discussed the synthesis protocol of CFMNPs. In brief, 3 mmol of iron acetyl acetonate and 1.5 mmol of cobalt acetate were dissolved firstly in a mixture of 40 mL phenyl ether and 10 mL of ethylene glycol. Next, 4mL of oleic acid and 3 gm of urea was added to this solution and followed by heating for 1 h at 160°C. We kept the samples at room temperature to cool down and after that washed with distilled water, ethanol and acetone. Next, the samples were dried and annealed at 300°C for 1 hr. Here, two sets of CFMNPs have been prepared by changing the oleic acid concentration only. 4 mL and 2 mL of oleic acids have been used to synthesize set A and set B particles respectively.

4.2.4. Synthesis of cobalt ferrite nanoparticle (CF NP):

The synthesis procedure of $CoFe_2O_4$ NP (CF NP) was discussed earlier. In brief, $FeCl_3.6H_2O$ & $CoCl_2.6H_2O$ were dissolved in distilled water with a molar ratio of 2:1 and mixed well. The solution was then heated and stirred and after reaching the temperature of this solution at 85^0 C, the KOH solution was added to it and continuous stirred for 1 h under heating. The sample was then allowed to cool at room temperature and washed with distilled water, alcohol & acetone. Next, the sample was separated by centrifugation and dried at air and finally stored in dry place.

4.2.5. Surface coating and Drug loading of the MNPs:

4.2.5.1. Loading of doxorubicin hydrochloride with Fe_3O_4 MNPs:

For drug loading, 40 mg of Fe_3O_4 MNPs was dispersed in 1 mL of de-ionized water into which 40 μL of ammonium hydroxide solution (1×10^{-3} M) was added and stirred for 30 minutes for proper mixing. Aqueous DOX solution was prepared by dissolving DOX with de-ionized water with concentration of the solution ~ 1.4×10^{-3} M. The DOX solution was added to the above solution under stirring condition and allowed to stir for another 30 minutes to ensure the drug to be loaded in the Fe_3O_4. Then the drug loaded particles were gently washed with de-ionized water to remove non-bound drug and dried. The drug loading and releasing efficiency has been checked by using UV-Vis absorption spectroscopy. The drug loading percentage for every experiment was calculated using the following formula represents below;

$$\textbf{Drug loading efficiency} = \frac{\text{Intensity}_{\text{Initial drug solution}} - \text{Intensity}_{\text{Final drug solution}}}{\text{Intensity}_{\text{Initial drug solution}}} \times 100\%$$

4.2.5.2. Folic acid coating with the CFMNPs:

For folic acid (FA) coating with CFMNPs, 10 mg of FA (97% purity) was dissolved in 10 mL of distilled water and mixed well to get a homogeneous mixture. After that, 10 mg of particles were added to this FA solution and stirred overnight in dark condition. Next day, the FA coated particles were separated from the solution by following the magnetic separation method and washed with water to remove excess unbound FA and finally dried overnight.

4.2.5.3. Loading of doxorubicin hydrochloride with FA coated CFMNPs:

At first, 10 mg of FA coated CFMNPs were taken in 5mL of distilled water and 100 μL of sodium bicarbonate (1×10^{-3} M) was added to this for making the particles water soluble. On other side, Dox solution was prepared by dissolving 6.897×10^{-4}M of doxorubicin hydrochloride (98%-102% purity) in de-ionized water. From the DOX solution, 100 μL solutions was added to the above particles solution and allowed to stir for 1 hr. Next, these DOX loaded FA coated CFMNPs were separated from solution and were gently washed with pH 7.4 PBS buffer and again dried properly for further use.

4.2.5.4. Loading of dopamine with $CoFe_2O_4$ NP:

The procedure for dopamine (DA) loading with $CoFe_2O_4$ nanoparticle (CF NP) is very easy process. At first, dopamine was dissolved in 2 mL of distilled water into which the CF NP was added. The solution was then continuing stirring for 3 h at room temperature to allow proper dispersion of nanoparticle and binding of CF NP with dopamine. This dopamine loaded CF NP solution was denotes as sol.A.

4.2.5.5. Functionalization of DA loaded CF NP with poly ethylene glycol:

The DA loaded CF NP needs to be functionalized with poly ethylene glycol (PEG) for proper delivering of dopamine inside the cell. After preparation of sol.A; sol.B and sol.C had been prepared. PEG_{4000} was dissolved in 3 mL of distilled water and sonicated for 5 minutes to make the sol.B. Here, we take PEG and DA added CF NP at concentration of 1:1. After that, sol.C had been prepared by dissolving Pluronic P123 in 500 μL of distilled water and sonicated this solution for proper mixing. Then the sol.B and sol.C was mixed to get a new mixture named sol.D. After 3 h of stirring, the sol.A was sonicated on a bath sonicator and at the time of sonication, the Sol.D was added to sol.A drop by drop with the help of 1 mL syringe. Then the mixture was further sonicated at room temperature for 1 h. After that, the mixture was again probe sonicated for 10 min. and kept overnight under stirring condition. Next day, the PEG coated CF-DA particles was separated by centrifugation and gently washed with distilled water and ethanol. At last, the final sample CF-DA-PEG was dried on a Lyophilizer.

4.2.6. Physicochemical Characterization:

The crystallinity and phase of the synthesized MNPs were characterized by XRD. The XRD measurements were done on powder samples at room temperature. All the XRD patterns were measured by using Rigaku Miniflex II diffractometer using Cu K_{α} (λ= 1.5418 Å) radiation source with incident angle 2θ. The morphology of the MNPs was investigated by transmission electron microscopy (TEM). TEM, High Resolution Transmission Electron Microscopy (HRTEM) of MNPs were performed in TECHNAI G^2 TF20 at 200 kV. For TEM analysis, samples were dispersed in alcohol and sonicated for proper mixing. The samples were then drop casted on 300-mesh carbon-coated copper (Cu) grid and dried in vacuum desiccators. By using Fourier Transform Infrared (FTIR) spectroscopy (NEXUS-470, Nicolet, USA), we have analyzed the chemical properties of this MNPs. KBr matrix were used to performed the FTIR where samples and potassium bromide were present with a mass ratio of 1:100.

4.2.7. Drug release studies:

The drug release efficiency from these MNPs was investigated by Shimadzu model UV-2600 spectrophotometer using quartz cuvette of 1 cm path length. For drug release study, the respective drug loaded MNPs were well dispersed in phosphate buffer solution (PBS) and the UV-Vis absorption spectra have been recorded in different condition such as at two different pH (pH 7.4 & pH 5) and temperatures (37^0C & ~45^0C) and also under AC magnetic field. Generally, the cancer cell pH is lower than normal cell pH and the temperature between 41^0C-45^0C denotes as hyperthermic temperature.

4.2.8. Cellular experiments:

4.2.8.1. Cell culture:

The human cancer cell lines i.e. lung cancer cell line A549 and HeLa cell line were purchased from National Centre for Cell Science (NCCS, Pune, India). The cells were maintained in different medium such as Dulbecco's Modified Eagle's Medium (DMEM), Roswell Park Memorial Institute (RPMI-1640) medium etc. containing 10% of fetal bovine serum (FBS) at 37^0C in 5% CO_2 incubator. To get off contaminations of cells, different antibiotics such as penicillin, streptomycin and gentamicin (100mg/L each) has been added to these medium.

4.2.8.2. Heating effect of drug loaded Fe_3O_4 MNPs on cells:

The A549 cells were seeded on 6-well plates and incubated for 24 h. Next day, the cells were treated with specific doses of DOX loaded Fe_3O_4 MNPs and incubated for 2 h at 43^0C under thermal environment. The cell plates were then again shifted back to 37^0C. Another set of control cells without treatment were incubated at 37^0C in a CO_2 incubator for 24 h. The HeLa cells were maintained in RPMI-1640 medium which were also treated with specific dose of DOX loaded Fe_3O_4 MNPs and incubated for 4 h at 43°C. After that, the aliquot of cell line solution was exposed with trypan blue dye from where 20 μL of solution was put on a glass slide and covered by cover slip. The images of cells were taken by using a high resolution optical microscope.

4.2.8.3. Cellular experiments for dopamine delivery:

4.2.8.3.1. Cell viability assay:

3(4,5-dimethylthiazolyl-2)2,5- diphenyl tetrazolium bromide assay (MTT assay) was performed according to manufactured protocol to check the cell viability or metabolic status of cells. The cells were seeded on 96-well plates and incubated for overnight at 37^0C in 5% CO_2. Next day, the cells were treated with different doses of samples and kept in incubation for further 24 h. The cells were then washed with PBS and suspended in colourless DMEM. After that, the cells were followed by further incubation for 6 h after adding of MTT solution. 100 μL of DMSO was then added to each well to dissolve the formazan crystals and kept in rocking condition for 10 minutes. At last, absorbance of the purple colour formazan was recorded at 570 nm by using a Bio-Rad (model 550) Elisa Micro plate reader. The given formula was used to calculate the percentage of cell viability ...

$$Cell\ viability = \frac{Number\ of\ living\ cells}{Initial\ number\ of\ living\ cells} \times 100$$

4.2.8.3.2. Cellular internalization of dopamine loaded MNPs:

Internalization of cobalt ferrite (CF) NP and PEG functionalized dopamine loaded CF NP (CF-DA-PEG) inside the cells was checked by labelling them with fluorescent dye RITC. For RITC labelling, 25 mg of particles were dissolved in 0.1 M $NaHCO_3$ solution and sonicated to get a homogenous mixture. 1 mg of RITC was dissolved in 2 mL of aqueous DMSO (1:1, v/v) and added this solution with $NaHCO_3$ made particles dispersions. The mixtures were then rocked in dark for 24 h at room temperature. Finally, the RITC labelled CF & CF-DA-PEG particles were culled by centrifugation and washed gently with water to remove untagged RITC.

4.2.8.3.3. Colony formation assay:

To check the sensitivity of dopamine inside the cell, in vitro colony formation assay was performed in A549 cell line. The cells were seeded on 6-well plates and incubated for 24 h upon treatment with different doses of CF-DA-PEG (2.5, 4, 6 and 8μg/mL). Next day, the cells were washed with PBS and re-suspended in fresh medium. The cells were then incubated at 37^0C in 5% CO_2 for 10 days which allow the viable cells to grow into colonies.

After 10 days, cells were washed with PBS and fixed in 3.7% formaldehyde solution followed by further PBS washing. Finally, the cells were stained with crystal violet and again washed with PBS for removing of excess stain. The colonies formed by A549 cells were observed under microscope (OLYMPUS).

4.2.8.3.4. Determination of cell cycle by flow cytometry:

For cell cycle analysis, the A549 cells were seeded with confluency of 2×10^5 cells/well and treated with different doses of CF-DA-PEG followed by overnight incubation at 37^0C. The cells were then trypsinized after PBS washing and collected by centrifugation. Then the cell pellets were fixed with 70% ethanol and stored at -20^0C for 3 min. The RNase A (2mg/mL) was then mixed with cells and incubated at 37^0C for 30 min followed by addition of 3 μg/mL of Propidium Iodide. After treatment, the changes of distribution of nuclear DNA in different cell cycle phase was determined by using a flow cytometer, equipped with 488 nm argon laser light source and 623 nm band pass filter (linear scale) and CellQuest software (Becton-Dickinson, San Jose). Ten thousand total events were obtained and the DNA content was plotted on histogram showing the PI fluorescence on x-axis and cell counts on y-axis.

4.2.8.3.5. Determination of cellular apoptosis of A549 cells by Annexin V:

After treatment with CF-DA-PEG, the cellular apoptosis were checked by flow cytometer. For this, A549 cells were seeded on 12 well plates with confluency of 2×10^5 cells/well and incubated for overnight upon treatment with different doses of CF-DA-PEG. Then the cells were washed well with PBS and followed by trypsinization. The cells were collected by centrifugation at 3000 rpm for 3 min and re-suspended in AnnexinV binding buffer containing 5mL/100mL of AnnexinV-FITC and 5mg/mL of PI followed by incubation at 37^0C for 15 min. Different apoptotic stages of cells such as early & late apoptosis; necrosis and cell death was determined by stating the cells with AnnexinV and PI and were analyzed by flow cytometer.

4.2.8.3.6. Measurement of intracellular ROS production in A549 cells:

Different sets of A549 cells were seeded on plates with density of approximately 2×10^6 cells per well. The cells were treated with different amount of CF-DA-PEG and incubated for 24 h. Next day, the cells were scrapped and were collected by centrifugation which operates at 300g for 5 min at room temperature. The cell pellets were then re-suspended in 1mL of pre warmed PBS to 37^{0}C and followed by addition of H_2DCFDA to each sets of sample in such a way that make the final concentration of 2 μM. The all sample sets were then incubated at 37^{0}C in dark for 20min. ROS generation by this particle in different doses was measured by using the flow cytometer (BD FACS, Verse BD Biosciences, San Diego, CA) at an excitation and emission wavelength of 488nm and 520 nm respectively using the FACSuite software.

4.2.8.3.7. JC-1 staining mitochondrial membrane potential:

Changes of mitochondrial depolarization in A549 cells after treatment with CF and CF-DA-PEG was observed by using a fluorescent cationic dye JC-1, obtained from JC-1 mitochondrial membrane potential detection kit (Biotium, Hayward, CA, USA). For this experiment, the cells were treated with specific doses of CF and CF-DA-PEG and incubated for 6hr and finally analyzed under fluorescence microscope (EVOS FL).

4.2.8.3.8. Bidirectional wound healing assay:

By wound (scratch) healing assay, we have checked the cell migration property of A549 cells after treating them with CF-DA-PEG. For this assay, the cells were seeded on 12 well plates with 85-90% confluency and treated with different doses of CF-DA-PEG (1.5 μg/mL & 2.5 μg/mL). A bidirectional wound had been created on cell plates by scratching the monolayer of cells by a sterile 10μL tip and incubated the cells for 24 h. After that, the cell migration was checked by using a camera (OLYMPUS) fitted phase contrast microscope. ImageJ software was used to determine the average migration distance which was covered by healthy cells. Here, migrations of the treated cells were compared with control cells where the migration rate of control cells was taken as 100%. The data were taken as triplicate manner and calculated as the means ± S.E.

4.2.8.3.9. Transwell Migration Assay:

Transwell Migration Assay was also performed to check the migration of cells. For this assay, the cell culture inserts (BD Biosciences, Sparks, USA) with pores size 8 mm were placed inside the 12-well plate containing 100 mL of DMEM without FBS. Then inside the chambers, about 2X10^5 cells were seeded in the upper half of the inserts where in the lower chamber of the 12 well plates, 10% FBS containing DMEM was added. Then the cells were followed by treatment and incubated for 24 h to allow cell growth. Next day, the cells present on the top of the inserts were washed with PBS and fixed with 3.7% formaldehyde. The cells were then permealized with methanol and stained by using Giemsa stain for 30 minutes. The cells which were present in the lower portion of the inserts were then counted in five microscopic fields per well and the extent of migration was expressed as an average number of cells per microscopic field.

4.2.8.3.10. Scanning electron microscopic image of A549 cells:

By using the scanning electron microscopy, we have observed the changes of A549 cell morphology after treatment with different doses of CF-DA-PEG. At first, the cells were cultured on cover slip in 1 mL plate for overnight at 37^0C in 5% CO_2 followed by treatment with specific doses of CF-DA-PEG and further incubated for 24h. Then, the cells were washed well with pre warmed PBS and fixed with 3.7% formaldehyde at room temperature for 20min and further washed with PBS. The cells containing cover slips were then dipped sequentially into various concentration of alcohol solution (10%, 30%, 50%, 70%, 90% and 100%) to allow dehydration of cells and dried in open air. Finally, the cells were observed under SEM (ZEISS, Germany)- EVO-18 special edition.

4.2.8.3.11. Western blot analysis:

A549 cells were seeded on a 6-well plate (2.6×10^5 cells/well) and incubated for a particular time at 37^0C upon treatment with CF-DA-PEG. After that, the cells were trypsinized and followed by treatment with lysis buffer (20 mM HEPES pH 7.5, 10 mM KCl, 1.5 mM MgCl2, 1 mM Na-EDTA, 1 mM Na-EGTA, and 1 mM DTT) in which protease and phosphatise inhibitors mixture was present. Protein estimation was calculated by using bicinchoninic acid (BCA) assay kit (Merck, Germany). In case of all sets of samples, an equal amount of proteins was separated by SDS-PAGE and then transferred on PVDF membrane (milipore) which were then incubated with primary antibodies of target gene such as, anti-p53, Bax, Bcl2, Caspase-3, Caspase-9 and Cyt-c with respect to the internal control COX IV and β-Actin (purchased from Santa Cruz Biotechnologies, USA). Finally, the ALP-conjugated secondary antibodies (USA) was used to develop the blots with NBT/BCIP (1:1) dissolved in ALP buffer.

4.2.8.3.12. RNA isolation and quantitative real time PCR:

Total amount of RNA had extracted from cells by using the TRIzol reagent (Invitrogen) following the manufactur's protocol. Reverse transcription of RNA was done by using oligo(dT)$_{18}$ primer (Biobharti, India) which provided in Transcription First Strand cDNA synthesis kit. The qRT-PCR was performed with 50 μg of CF-DA-PEG sample on Light cycler 96 (Roche, Germany) by using SYBR green ready mix (Roche, Germany) and specific primers. Glyceraldehydes-3-phosphate dehydrogenase (GAPDH) or 18S were used for normalisation of mRNA quantification. Gene expressions were defined from the threshold cycle (C_q), and the $2^{-\Delta Cq}$ method was used to calculate the relative expression levels after normalization with reference to expression of housekeeping genes (GAPDH or 18S).

4.2.8.3.13. Fluorescence Imaging of A549 cells:

For fluorescence imaging, the cells were seeded on cover slips in 1 ml plates and treated with specific doses of CF-DA-PEG followed by incubation for 12 h at 37^0C. After that, the cells were fixed with 3.7% formaldehyde and permealized with 0.1% Triton X-100. Then the cells were treated with 5% BSA in PBS buffer and incubated with primary antibodies such as anti cytochrome-c, anti P^{53} followed by FITC conjugated secondary antibody (Sigma). Here, for staining of cell nucleus, DAPI (Invitrogen, Carlsbad, CA) had been used. At last, the samples were mounted on glass slides with Prolong Gold Antifade Reagent (Invitrogen, Carlsbad, CA) and observed under fluorescence microscope (EVOS FL).

4.3. Results and Discussions:

4.3.1. Stimuli dependent drug (doxorubicin) release by cube shaped magnetite nanoparticles:-

4.3.1.1. Structural and morphological analysis of magnetite nanoparticles:

The XRD pattern of the synthesized Fe_3O_4 MNPs is represented at Fig.4.1. All the diffraction peaks were matches with the JCPDS card no. 19-629 of Fe_3O_4 MNPs which has cubic inverse spinel structure with a face-centered cubic (fcc) unit cell composed of 32 Oxygen anions, 16 Fe(III) cations and 8 Fe(II) cations. Among all the cations, half of the Fe(III) cations are coordinated tetrahedrally while the other half and all of the Fe(II) cations are coordinated octahedrally [47,48]. For confirming the magnetite phase, the lattice parameters were calculated using the XRD peaks at 511, 440 and 533 planes which occur at 2Θ value of 56.93, 62.57 and 74.1078 respectively. Here, the high angle peaks are considered to minimize the error. The lattice parameters obtained for XRD peaks at 511, 440 and 533 planes are

8.401, 8.397 and 8.398Å respectively. Among these, we take the average lattice parameter i.e. 8.398Å which matches with the lattice parameter of magnetite. So, it can be concluded that the synthesized particles are of pure phase magnetite. The average crystallite size of this particle was calculated from the 311 peak using Debye Scherrer's equation, which was found to be about 18 nm.

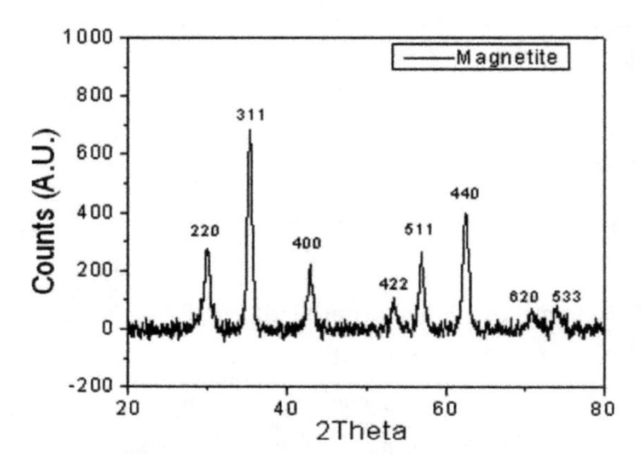

Fig.4.1. The XRD pattern of Fe_3O_4 MNPs.

The TEM images are represents in Fig.4.2. The TEM image also confirms that the particles are of cube like shape with average size of 18±1.5 nm. From this image, some of the particles have been observed are hexagonal which is due to tilt the image in different angles during TEM image analysis. Generally, TEM image gives two dimensional views, so, from other view angle, the cube shaped particles may look like hexagonal, rectangular, square etc. are shown on top of the Fig.4.2.(a,b,c,d). The high resolution transmission electron microscopy (HRTEM) (Fig.4.2.(f)) was performed to see the inter-planer distance between the lattice fringes is calculated to be about 0.252 nm that correspond to the distance between 311 planes of the Fe_3O_4 crystal lattice. Fig.4.2.(e) represents the electron diffraction pattern of Fe_3O_4 MNPs and analysis of this pattern again confirms that the particles are pure phase magnetite.

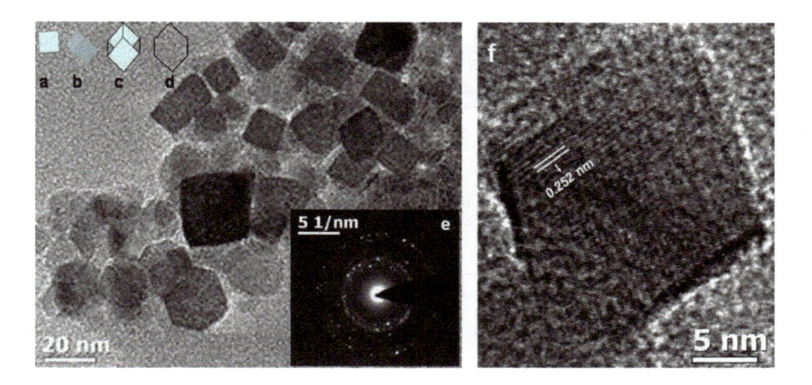

Fig.4. 2. TEM image of Fe_3O_4 MNPs (a), (b), (c), and (d) show the pictures of the particles from different viewing angles; (e) the electron diffraction pattern of this particle, (f) the HRTEM image of Fe_3O_4.

4.3.1.2. AC magnetic properties of Fe_3O_4 MNPs:

A lab made set-up has been used to perform the AC magnetic measurements of cube shaped Fe_3O_4 MNPs where we can produce maximum ~ 90kA/m AC magnetic field at nearly 700 Hz frequency. Fig.4.3.(a) shows the AC hysteresis loops of the Fe_3O_4 MNPs measured at room temperature in the frequency ranges from 50 Hz to 600 Hz. Here, we have noticed that the hysteresis loop area of the cube shaped Fe_3O_4 MNPs are increases with increasing the frequency of AC field. Generally, the magnetic nanoparticles behave like either superparamagnetically or ferromagnetically depending on frequency. Here, the Fe_3O_4 MNPs are behaves like ferromagnetic. Hergt et al. demonstrated that under an AC magnetic field, the magnetic nanoaprticles show remarkable heating effects which related to losses during the magnetization reversal process of these particles, defined as specific loss power (SLP) [49]. In general, the temperature enhancement occurs in magnetic nanoparticles when they are influenced by external high frequency magnetic field. Fig.4.3.(b) represents the variation of SLP (proportional to the product of hysteresis loop area and frequency) with AC field frequency. Here, the SLP of the Fe_3O_4 MNPs increases with increasing frequency of applied AC field. We have measured the SLP of Fe_3O_4 up to 600 Hz AC field because after that the impedance of the lab set up circuit becomes so high that the reduced field is not enough to saturate the hysteresis loops. We are interested to use this particle for drug release triggered by AC magnetic field hyperthermia therapy. So, we have measured the heat induced capacity of this particle in solution under different applied AC frequencies and observed that 10 mg of

Fe_3O_4 when dissolved in 3 mL of solution, it produced 43^0C heat in the solution at 600 Hz applied AC frequency.

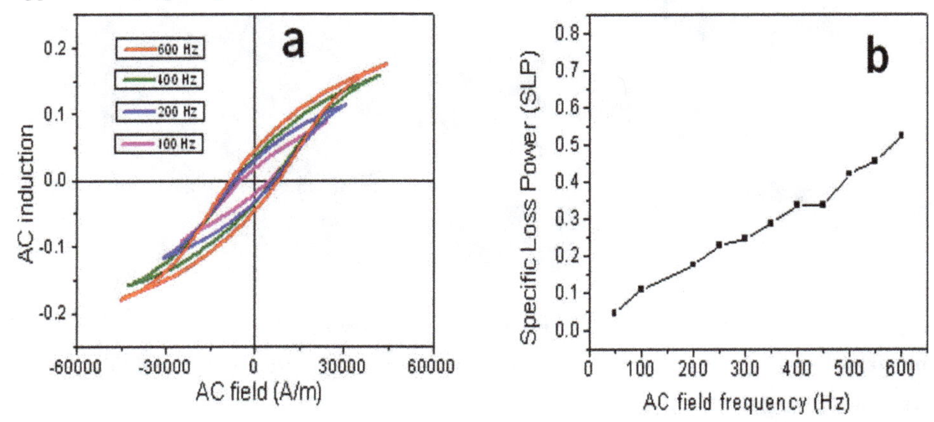

Fig.4.3. (a) Hysteresis loops of Fe_3O_4 MNPs at different AC field frequenciesand (b) Specific loss power at different AC field frequency.

4.3.1.3. Different stimuli dependent drug release studies:

By using the UV-visible absorption spectra, the amount of DOX loaded with Fe_3O_4 MNPs was calculated from the absorption spectra at 500 nm of initial DOX solution which was take after washing the drug loaded particle. The loading efficiency of our particle is 72% which is much better than previous published result [50]. During the drug loading procedure, functionalization of Fe_3O_4 with NH_4OH may increases the loading efficiency of drug with particle because of OH^- group on particle's surface. It is known that the surface area of cubic shaped particles is generally higher than the same volume of spherical shaped particles. So, this can be the other reason for better drug loading with the Fe_3O_4 MNPs due to its cubic shape. We have performed the drug release experiments at different condition. 10 mg of DOX loaded Fe_3O_4 MNPs were dissolved in 3 mL of PBS solution and the drug release was recorded at certain interval at two different pH (5 & 7.4) and temperature (37°C & 43°C) (Fig.4.4.(a)). From Fig.4.4.(a), it was observed that at lower pH (pH5) and higher temperature (43^0C) the drug release rate is very high compared to other condition. Another experiment was done to investigate the sensitivity of drug loaded Fe_3O_4 at lower pH. For this, 100 mL of pH~5 buffer solution had been prepared following the proper method. Small amount of Fe_3O_4 MNPs were dissolved in 2 mL of this buffer solution and the UV-Vis absorption spectrum of this sample containing buffer solution was recorded at zero time. After that, several absorption spectra were recorded at every 10 min intervals. On other side, small amount of Fe(III) ion was dissolved in 2 mL of above mentioned pH~5 buffer solution and the absorption spectra of this solution was taken. Fig.4.4.(b) represents all the UV-Vis absorption spectra and observed that all the peaks are generated at 312 nm (λ_{max}) and the intensity of UV absorbance value of buffer containing Fe_3O_4 particles was increased with increasing time. Hence, at lower pH, the Fe_3O_4 MNPs gets dissolute and free Fe(III) ions are formed and the dissolution rate was proportional to time. Therefore, during loading, the drugs bind with the oxygen molecule of the Fe_3O_4 MNPs is presented in a schematic diagram shown in Fig.4.5. At lower pH, the MNPs gets dissolved and the drugs getting detached from the particles [51]. Generally, the cancer cells is more acidic compared to normal cells and the pH of cancer cell regions remain in the range of 5 to 5.5 [52,53]. So, use of the DOX loaded Fe_3O_4 MNPs for drug delivery is more significant and shows high potentiality for cancer therapy. On other side, better release of drug has been observed at 43^{0}C than 37^{0}C. Due to thermal agitation at higher temperature, the drug loaded particles becomes vibrated which breaks the bonds between drugs and particles and facilitates the drug release. So, by this method the drug release property can be enhanced compared to others [54-56]. Our particles also produced heat under AC magnetic field and this heating can be controlled by changing the frequency of AC magnetic field. So, the drug release can be controlled noninvasively by changing the AC field frequency and by this method, the pulsed release of drugs also possible accordant to requirement.

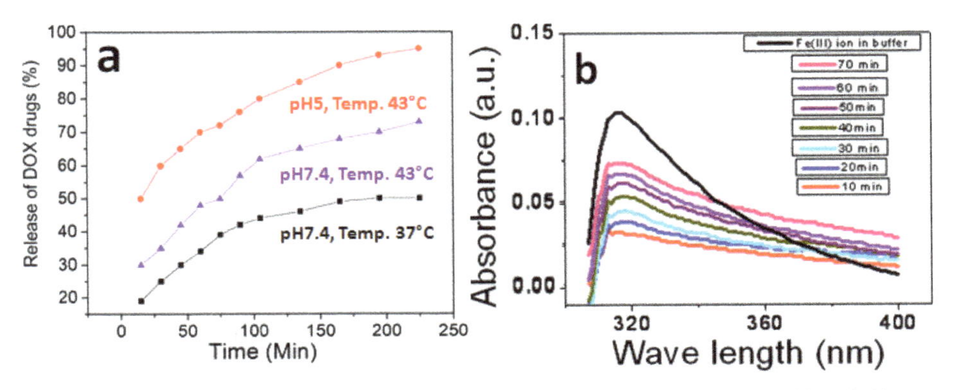

Fig.4.4. (a) release of DOX by Fe_3O_4 MNPs at different temperatures and pH and (b) UV-visible absorption spectra of Fe_3O_4 MNPs in low pH buffer solution measured at 10

min interval of time under stirring condition and of Fe(III) ions dissolved in the same buffer solution.

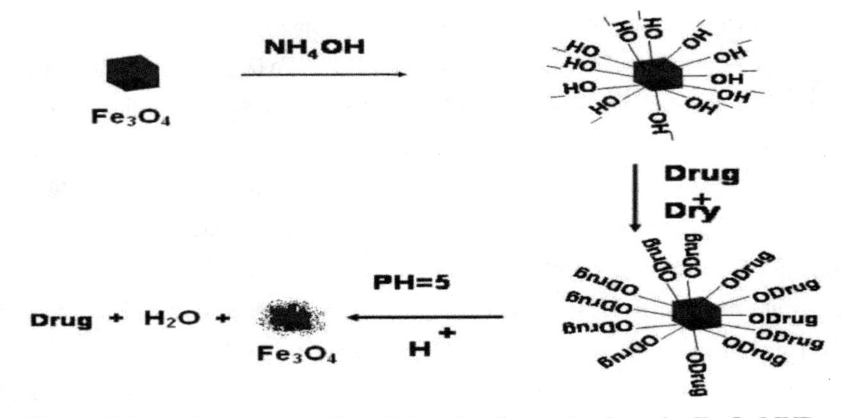

Fig.4.5. Schematic representation of drug loading and release by Fe_3O_4 MNPs.

4.3.1.4. In-vitro studies:

The heating effect of DOX loaded Fe_3O_4 MNPs on A549 cells was observed and compared with the untreated cells. The transmitted images were taken by using EVOS-FL microscope (Invitrogen, USA) represents in Fig.4.6.(a&b). Here, we have noticed that, after treatment with drug loaded particles at 43^0C, most of the cells become dead and they starts to float (Fig.4.6.(a)) but almost all cells remain alive in case of control (Fig. 4.6.(b)). The death rate of cancer cells was also checked on HeLa cell line where the HeLa cells were treated with DOX loaded Fe_3O_4 MNPs and incubated at 43^0C. Here, the aliquot of cell line solution was further treated with trypan blue dye and the image was taken in a high resolution optical microscope (Fig. 4.6.(c)). Higher cell death has been observed here. So, we are able to kill the cancer cells by using DOX loaded Fe_3O_4 MNPs at high temperature. It is known that DOX is a cytotoxic drug that is generally administered in the body of cancer patients through intravenous, intra-vesicle and intra-arterial routes. DOX is used for cancer therapy which works by intercalating DNA with the most serious side effects such as damaging of heart, liver etc [57,58]. This drug has been used to treat wide range of cancers like haematological malignancies (leukaemia, lymphoma), soft tissue sarcomas, malignant tumours etc. So, by using our method, controlled release of this drug in desired location can be achieved without spreading it all over the body by applying AC magnetic field of suitable amplitude and frequency from outside the body. The efficiency of drug release by our proposed method is also be enhanced by pH with AC magnetic field.

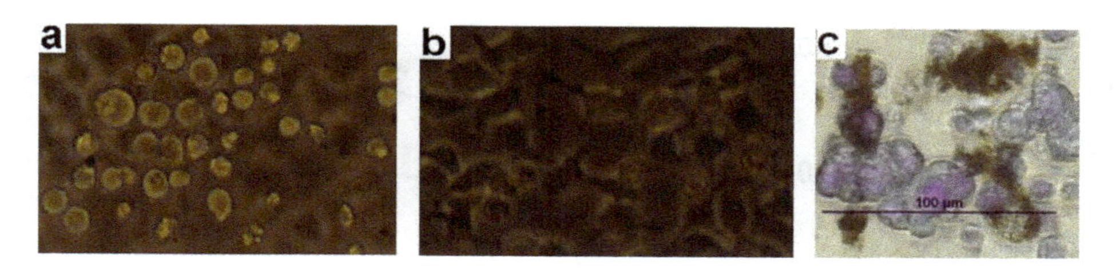

Fig.4.6. (a) Transmitted images of cancer cells after treatment with the DOX loaded Fe_3O_4 MNPs at dose of 150 µg/mL at 43°C, (b) control and (c) optical microscopy image of trypan blue dye treated cells after treatment with DOX loaded Fe_3O_4 MNPs at 43°C.

4.3.2. AC magnetic field dependent drug (doxorubicin) release by FA coated spherical CFMNPs:-

4.3.2.1. Structural and morphological analysis:

Details structural and morphological properties and AC & DC magnetic properties of these CFMNPs have already discussed in chapter 3. So, here we focused only on the drug release studies under AC magnetic field.

4.3.2.2. AC magnetic field dependent drug release studies:

We have performed the drug release studies for both sets of the particles (size of 250 nm and 350 nm). 10 mg of both DOX loaded FA coated particles were dissolved separately in 3 mL of PBS buffer solution and the intensity of drug release for the two sets of particles at different time interval has been recorded under AC magnetic field with the help of UV-vis spectrometer. Fig.4.7. shows the percentage of drug release with time for the two samples under maximum field of 43 KA/m at a 50 Hz frequency. It is observed that both the samples show efficient drug release under AC magnetic field but the drug release rate for bigger particles (350 nm) is higher than that of smaller particles (250 nm). Under AC magnetic field, the particles are thermally agitated due to hysteresis loss which causes detaching of the drug molecules from the particles and releasing of drug. The AC hysteresis loop area of the bigger particles is higher compare to smaller particles which may causes the higher heating ability of the particles and hence higher rate of drug release. Therefore, both sets of particles have the efficiency to be utilized for drug release by magnetic hyperthermia technique.

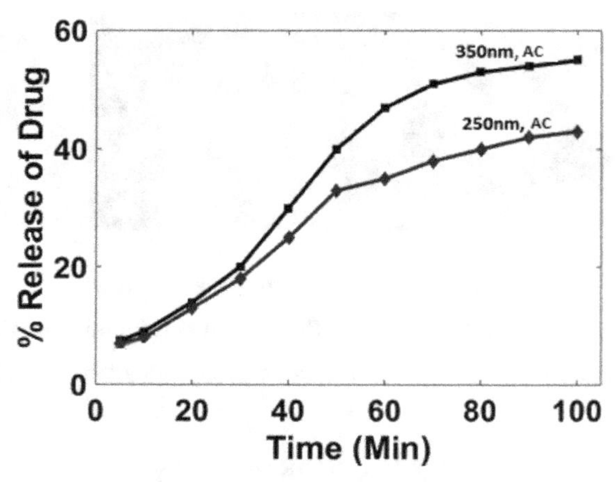

Fig.4.7.Time dependent AC magnetic field triggered drug release curves of 250 nm and 350 nm sizes of CFMNPs.

4.3.3. Delivery of dopamine by PEG functionalized cobalt ferrite nanoparticles for the treatment of non small cell lung cancer (NSCLC):

4.3.3.1. Structural and morphological analysis:

The X-ray diffraction patterns of the synthesized cobalt ferrite (CF) nanoparticles have shown in Fig.4.8. The diffraction peaks in the XRD spectrum were matches with the JCPDS data (ID: 22-1086) and confirmed that the synthesized nanoparticles is CF and are crystalline in nature. The CF NPs showing peaks at different 2Θ planes such as (220), (311), (400), (422), (511), (440) and (533). The average crystallite size of CF NP is calculated using the Debye Scherrer equation and found to be ~12.6 nm using (311) peak.

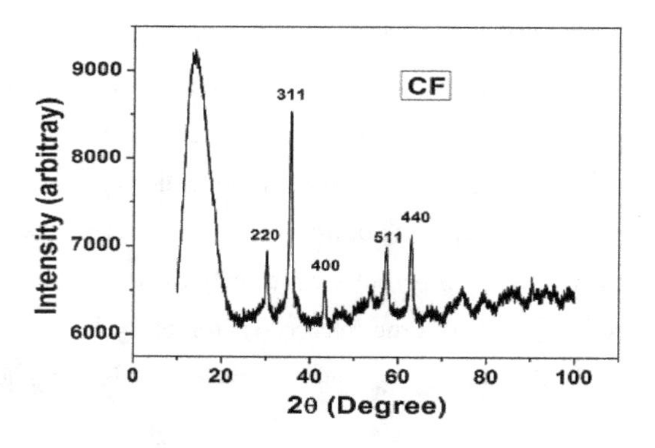

Fig.4.8. XRD spectrum of cobalt ferrite (CF) nanoparticles.

Fig.4.9. represents the FTIR spectra of CF, DA, PEG & CF-DA-PEG. In the dopamine spectrum, few peaks has been observed ranges from 3500 $cm^{-1}$to 2800 cm^{-1} are assigned due to stretching vibrations of the O-H, C-N and N-H groups of dopamine, respectively. Another small peaks have noticed in the dopamine spectrum ranges between 2778–2170 cm^{-1}, corresponding to different CH vibrations of either aryl or aliphatic CH bonds. The absorption bands around 1650 and 1500 cm^{-1} confirms the presence of aromatic C=C, C-N bonds in the DA, ensuring the presence of aromatic benzene ring and amine species in the DA. The peaks observed between 1494 cm^{-1} to 900 cm^{-1}, analogous to the bending vibration of CH group and the aryl oxygen stretching vibration, respectively [59]. In case of PEG molecule, the absorption band at ~3500 cm^{-1} corresponds to OH stretching and 2878 cm^{-1} is due to C-H stretching where the bands at 1464 and 1343 cm^{-1} occurs due to bending of C-H bond have been observed. Some more absorption bands have observed which occur due to O-H and C-OH bending vibrations at about 1200 to 1000 cm^{-1} [60]. In case of PEG coated DA loaded CF particles; we have seen some very little broad band at ~3500 cm^{-1} and 2800 cm^{-1} which are due to the stretching of N–H and O–H group, coming from both DA and PEG. Due to strong binding of DA and PEG with particles by these two bonds, the N-H and O-H peaks are becomes weaken and almost going to be vanished. Some peaks analogous to aromatic C=C, C-N bonds are coming from the DA and the peaks below this range are coming from both DA and PEG which are becoming less intense with broadening and shifting of the bands indicating the loading of DA with the particles and functionalized with PEG.

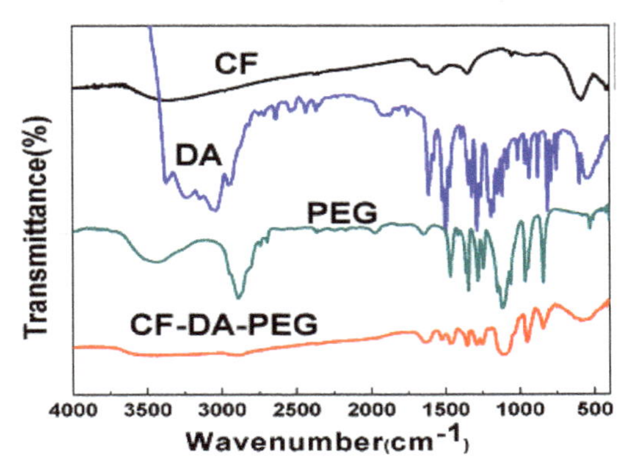

Fig.4.9. FTIR spectra of cobalt ferrite (CF), dopamine (DA), polyethelene glycol (PEG) and CF-DA-PEG.

The size and shape of CF & CF-DA-PEG were detected by TEM images represented in Fig.4.10. and observed that the average size of individual cobalt ferrite NPs is around 13 nm which supports the XRD Scherer analysis. Due to functionalize of DA loaded CF with PEG, it forms a coating around the CF-DA and form CF-DA-PEG shown in Fig.4.10. After PEG coating, the size of CF becomes increased.

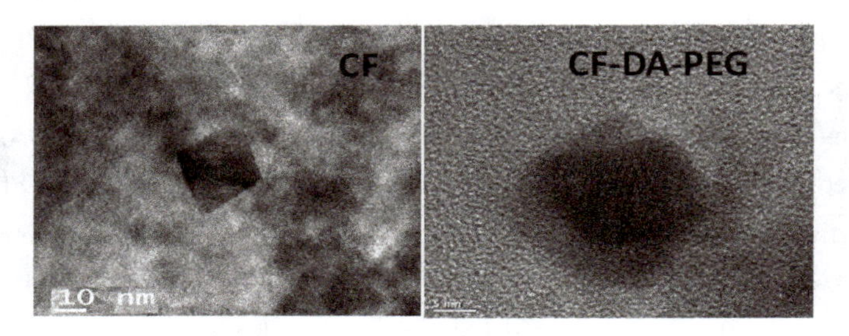

Fig.4.10. TEM image of CF and CF-DA-PEG.

We have checked the release of DA from CF-DA-PEG by using UV-Vis absorption spectra at two different pH (7.4 & 5) and temperatures (37^0C & 45^0C) are represented in Fig.4.11. The release rate of DA from CF-DA-PEG is very low at 37^0C and pH 7.4 compared to 45^0C at same pH. But higher release of DA has been noticed at pH 5 compare to pH 7.4 where the highest DA was released at higher temperature i.e. 45^0C and at pH 5. It is known that the cancer cells pH is about 5 which is more acidic than our normal cell pH i.e. pH 7.4. Hence, in cancer cells the dopamine release from this particle might be better compare to normal cells and provide a better drug release.

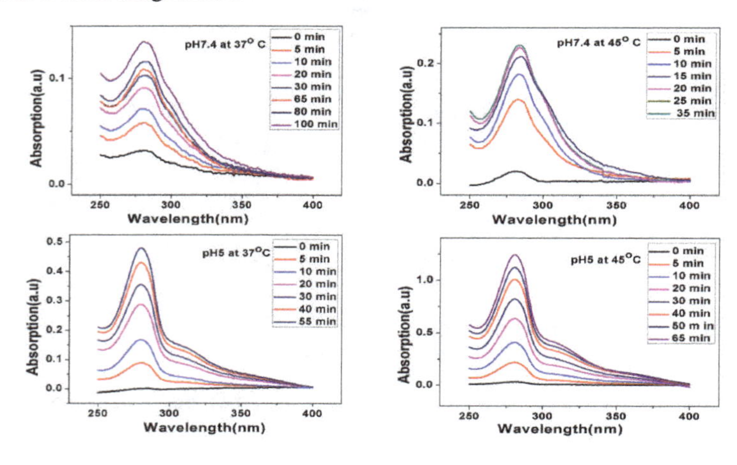

Fig.4.11. UV-VIS absorption spectroscopic data of releasing of DA from CF-DA-PEG at two different pH (pH7.4 & pH5) and temperature (37°C & 45°C).

4.3.3.2. Details cellular studies:

4.3.3.2.1. In-vitro cytotoxic study:

By MTT assay, we have checked the cytotoxicity of CF, DA, CF-DA & CF-DA-PEG on A549 cells are shown in Fig.4.12.(a&b) Here, dopamine shows the cytotoxic effects on NSCLC (non small cell lung cancer) cells and at dose of 10 μg/mL almost 50% cell death occurs (Fig.4.12.(a)). On other side, the CF has a less cytotoxic effects on this cell line where the CF-DA-PEG shows greater cytotoxicity and causes almost 57% cell death when CF-DA causes only about 18% death after 24 h of incubation at concentration of 5μg/mL (Fig.4.12.(b)). Here, the rate of cell death occurs in a dose dependent manner. Since, DA is highly water soluble, the CF-DA was coated with PEG and CF-DA-PEG have been prepared to stabilize the DA after binding with CF NPs. Coating of CF-DA with PEG increases the particle's biocompatibility, stability and promotes sustain release of DA into the cell cytoplasm. PEG gives a protection of the DA loaded particle before it reaches to the cells and thus PEG coating CF-DA particle becomes more effective to kill cancer cells and acts as a good drug carrier. In case of CF-DA, most of the DA gets released from the particle during making its dispersion and very little amount of DA remains inside the CF which actually is less effective to kill the cancer cells. Hence, to protect the DA inside the particle, the CF-DA has been coated by PEG which gives better result. DA is a neurotransmitter, naturally synthesized in our body and at certain concentrations; it has no adverse effects on normal body cells but our particles CF-DA-PEG shows anticancer effects on A549 cells. Generally, PEG attached with the surface of cells by formation of hydrogel which is homogeneously dispersed on cells and forms a good cell-adhesive [61]. So PEG coating helps the particles to easy deliver of the DA into the cells. In contact with cancer cells, more dopamine has been released due to acidic environment of cancer cells. From previous studies, researchers have seen that dopamine prevented the growth of blood vessels without causing many of the serious side effects [62]. Dopamine blocks the growth of new blood vessels in tumors by inhibiting the vascular endothelial growth which one of the causes of cancer cell death.

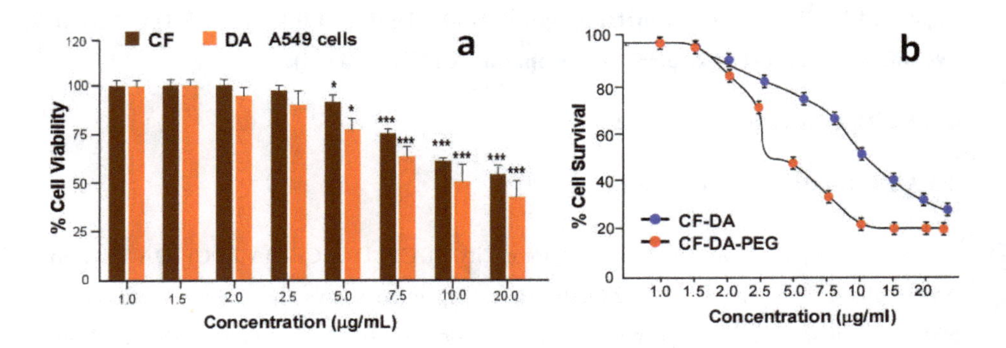

Fig.4.12. (a) Graphical representation of MTT assay showing percentage of A549 cell viability after treatment with different doses of CF & DA for 24 h incubation and (b) MTT assay showing comparative study of percentage of cell viability of A549 cells with different concentration of doses of CF-DA & CF-DA-PEG after 24h incubation.

Then we have checked the percentage of cell viability with respect to various times is represents in Fig.4.13. Here, we have seen that after treatment with both CF-DA and CF-DA-PEG, the cell survival rate was markedly decreased with respect to increase of incubation time but the percentage of cell death is much greater in case of CF-DA-PEG than CF-DA.

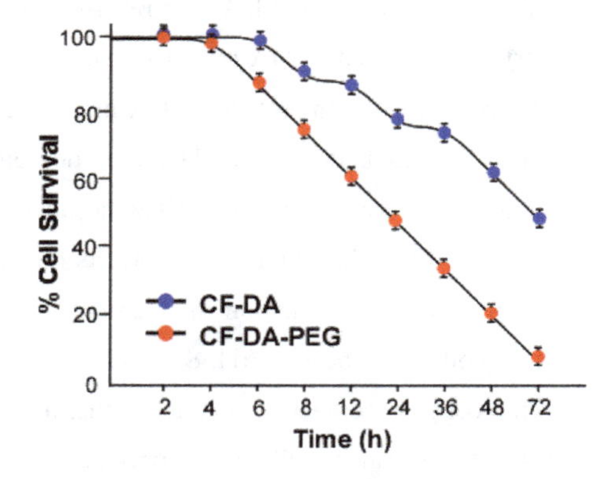

Fig.4.13. Percentage of cell viability of A549 cells at various incubation time (2,4,6,8,12,24,36,48 and 72 h) after treatment with equal concentration dose of CF-DA & CF-DA-PEG.

4.3.3.2.2. In-vitro cellular uptake study:

To check the cellular uptake of both CF & CF-DA-PEG into the A549 cells, these particles have been labelled with RITC which is used as a fluorescent marker. Cellular uptake of RITC labelled CF & CF-DA-PEG into cells after 24 h of incubation is shown in Fig.4.14. Here, it has observed that, more CF- DA-PEG enters into the cell compare to CF and more intense fluorescent comes out from the cytoplasm of CF-DA-PEG treated cells which again authenticates the cell adhesive property of PEG. Herein the cell nucleus has been stained with Hoechst-123 dye (Invitrogen, Carlsbad,CA). Fig.4.14. (right panel) represents the graph showing percentage of cellular uptake of CF & CF-DA-PEG in A549 cells at different concentration.

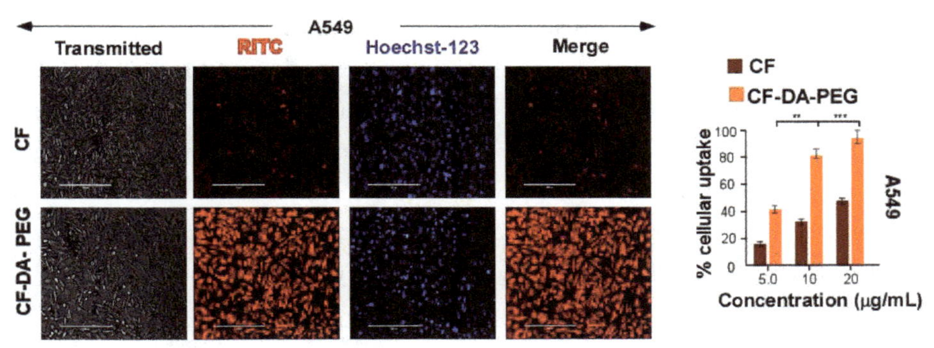

Fig.4.14. Fluorescence image of A549 cells after 24 h of incubation upon treatment with CF & CF-DA-PEG showing internalization of nanoparticles inside the cells (left pannel). From left to right images are indicating transmitted, RITC loaded nanoparticles, Hoechst-123 stained blue nucleus and merged images respectively. Right panel image shows the graphical representation of percentage of cellular uptake of RITC tagged CF & CF-DA-PEG in A549 cells at different concentration of treatment.

4.3.3.2.3. Effects of CF-DA-PEG on colony formation:

We have performed the clonogenic A549 cell survival assay upon treatment with different doses of CF-DA-PEG is represents in Fig.4.15. We have stained the colonies with crystal violet and observed that more cells have been stained with crystal violet in control because of more colonies have been formed in case of control set and the colony formation is gradually decreased in case of treated cells with increase the amount of doses. Therefore, the CF-DA-PEG significantly inhibits the colony formation of A549 cells in a dose dependent manner.

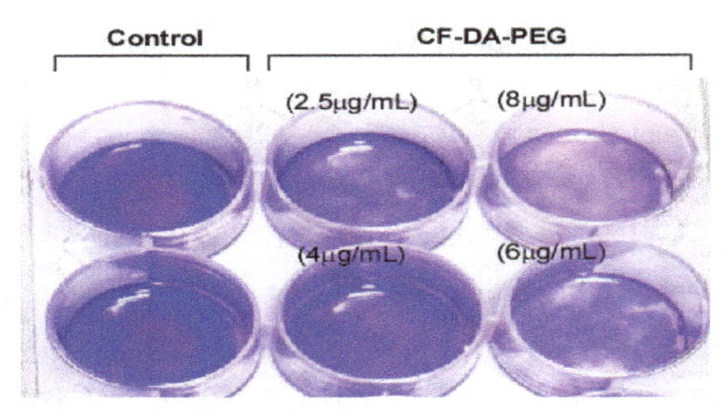

Fig.4.15. Colony formation assay of A549 cells after treatment with different doses of CF-DA-PEG.

4.3.3.2.4. Phase contrast microscopic images of treated A549 cells:

After that, we have taken some phase contrast microscopic images of CF & CF-DA-PEG treated A549 cell lines to re-establish the MTT assay results (Fig.4.16.). Due to treatment with CF & CF-DA-PEG, the cell to cell contact was markedly decreased. Significant changes in cell morphology and more cell death have been noticed after treatment with CF-DA-PEG compared to CF at same concentration. This result also supports the efficacy of CF-DA-PEG to deliver DA inside the cell which showing an anticancer effects on A549 cell lines.

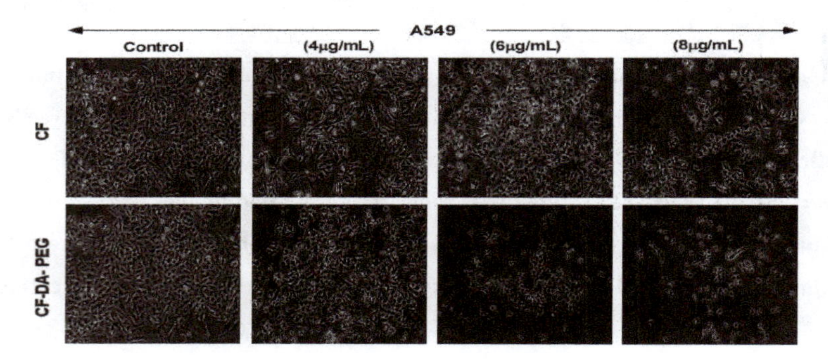

Fig.4.16. Phase contrast microscopic images of CF & CF-DA-PEG treated A549 cells after 24 h of incubation showing cell death at different doses.

4.3.3.2.5. Mitochondrial membrane potential analysis after treatment:

In addition, we have checked the effect of CF & CF-DA-PEG on mitochondrial function, since mitochondria play an important role in apoptosis. To monitor the mitochondrial membrane potential, JC-1 has been used basically which evaluate mitochondrial viability and function [63,64]. CF-DA-PEG cause's disruption of mitochondrial membrane potential in greater extent after 6 h of incubation compared to control and CF treated cells is shown in Fig.4.17. Here it has been observed that upon treatment with CF & CF-DA-PEG, the intensity of JC-1 green staining is increased & red fluorescence intensity is decreased and this change is more prominent in case of CF-DA-PEG treated cells indicating that CF-DA-PEG causes mitochondrial dysfunction in A549 cells with a disruption of membrane potential. The JC-1 is a cationic dye (monomer shows green fluorescence) which is capable to enter into the mitochondria by binding with the negative charge mitochondrial membrane, where it accumulates and starts forming dimer. This dimer emits red fluorescence in the maximum region at ~590 nm. So, in healthy cells with a normal membrane potential, the JC-1 dye enters into the mitochondria and accumulates there to form dimer which emits the red fluorescence. But in contrast, in unhealthy or apoptotic cells, lesser amount of the JC-1 dye can enters into the mitochondria because the inside of the mitochondria remains less negative. Under this condition, sufficient concentration of JC-1 does not enter for the formation of dimer or aggregates thus retaining its original green fluorescence. In our experiment, we have seen more intense green fluorescence in treated cells which means most of the cells are in apoptotic stage.

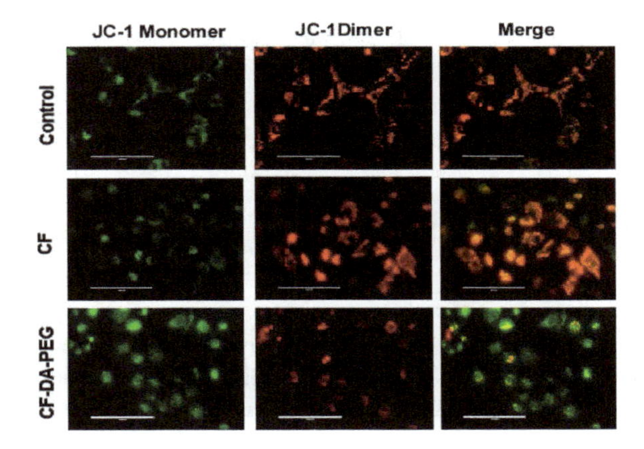

Fig.4.17. Fluorescence microscopic images of JC-1 staining A549 cells detecting the changes of mitochondrial membrane potential after treatment with CF and CF-DA-PEG compared with control.

4.3.3.2.6. Effects of CF-DA-PEG to produce ROS in A549 cells:

Under a physiological condition, the intracellular DA is prone to iron-catalyzed oxidation above certain concentration [65]. In our case, the CF particles consists of iron ions which help in oxidative catalysis procedure and the oxidation products coming out from this process are highly reactive and can produce ROS [66]. We have checked the ROS generation of A549 cells after treatment with different concentrated doses of CF-DA-PEG by using fluorescent dye DCFH-DA that measures hydroxyl, peroxyl and other reactive oxygen species activity within the cell and analyzed by flow cytometry represents in Fig.4.18. Here, the data is represented by histogram overlay where the fluorescence intensity of DCFDA was plotted on X-axis and cell count on Y-axis. Due to treatment with CF-DA-PEG, the shifting of curve has been observed towards right side compared to control which means increasing of fluorescence intensity of DCFDA is in treated cell than control. Hence, our particle CF-DA-PEG generates ROS and more free radicals inside the A549 cells. This ROS inhibits the cell proliferation & causes apoptosis of cells [67].

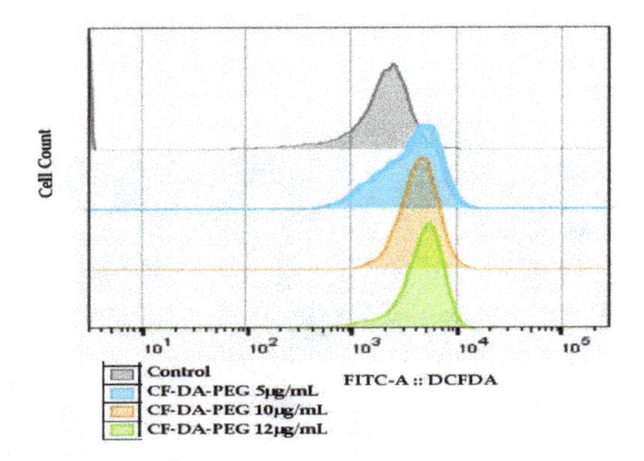

Fig.4.18.The histogram overlay represents the flow cytometric data of evaluation of ROS generation in CF-DA-PEG treated A549 cells at different doses measuring the DCFDA-fluorescence intensity.

4.3.3.2.7. Effects of CF-DA-PEG to induce apoptosis in A549 cells:

Then we have checked different stages of apoptosis of the cells by several experiments are represented in Fig.4.19. and Fig.4.20. Fig.4.19. shows the fluorescence microscopic images of AO/EtBr stained A549 cells after treatment with different doses of CF-DA-PEG and control. The dual acridine orange/ethidium bromide (AO/EtBr) is a type of fluorescent stain which is used to visualize the apoptosis associated change of cell membrane during the process of apoptosis [68]. By using this method, cells can also be distinguish according to different stages of apoptosis [69,70]. In general, granular yellow- green acridine orange nuclear staining indicates that the cells are in early stage of apoptosis and the cells marked with orange nuclear ethidium bromide staining means that the cells in late stage of apoptosis where the necrotic cells showed orange-red fluorescence at their periphery [71]. In our case, we have seen that the CF-DA-PEG treated cells at first go through early apoptosis at the dose of 5μg/mL to late apoptosis pathway at dose of 10μg/mL and at higher dose (12μg/mL) necrosis occurs (Fig.4.19.).

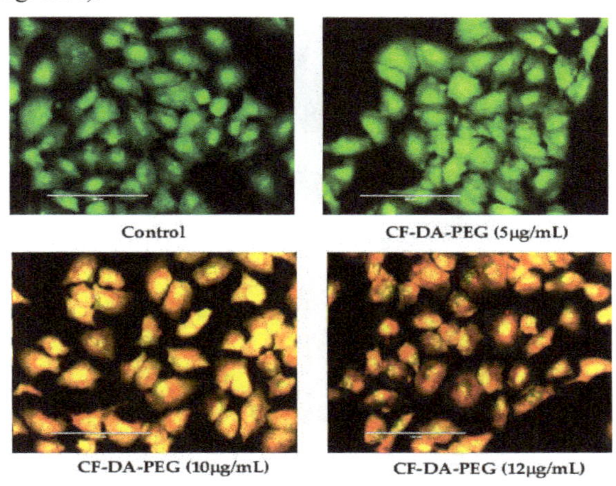

Fig.4.19. Fluorescence microscopy images of AO/EtBr stained A549 cells at different doses of CF-DA-PEG compare to control.

The cell apoptosis of CF-DA-PEG treated A549 cells was also investigated by flow cytometry where we have counted the number of Annexin-V-positive cells at different doses. The early and late apoptotic stages of cells are detected by stating the cells with conjugation of Annexin V and propidium iodide (PI). Generally, the intact membranes of healthy cells expel the PI, whereas the membranes of dead and damaged cells are permeable to PI. Therefore, the healthy cells are negative to both Annexin V and PI (Annexin V–/PI–), while cells that are in early apoptosis are positive to Annexin V and negative to PI (Annexin V+/PI–), and cells that are in late apoptosis or already dead are positive to both Annexin V and PI (Annexin V+/PI+) [72]. In our case, we have stained the untreated and CF-DA-PEG treated A549 cells using the Annexin V FITC apoptosis detection kit and the data was plotted against Annexin V FITC-A vs Propidium Iodide-A (Fig.4.20.), showing the cells that are presents in the non-apoptosis and various stages of apoptosis indicating the populations corresponding to viable and apoptotic cells. In Fig.4.20., the Q4 denotes the Annexin V–/PI–, Q3 Annexin V+/PI–, Q2 Annexin V+/PI+ and Q1 Annexin V–/PI+. Here, we have seen that, almost 96% cells are viable in control set which are remain in Q4 and percentage of viable cells decreases with increase of the amount of CF-DA-PEG. Upon treatment with CF-DA-PEG, the A549 cells go through early apoptosis to late apoptosis process depending on dose amount and at highest dose (12μg/mL) almost 42% cells are in non- apoptotic, 53% cells are in early apoptotic and 4% cells in late apoptotic conditions.

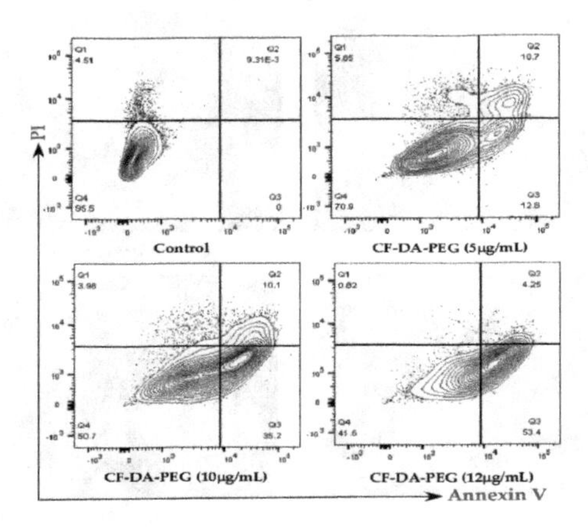

Fig.4.20. Annexin V/PI assay showing dose dependent apoptotic cell death of A549 cells after treatment with CF-DA-PEG analyzed by flow cytometry.

4.3.3.2.8. Analysis of cell cycle of A549 cells after treatment with CF-DA-PEG:

Cellular DNA content has been measured and the cell cycle was analyzed by flow cytometry.[73] From the flow cytometry data we got the percentage of the cell population in different stages of cell cycle such as G0/G1, S, and G2/M phases after staining the DNA with propidium idodide (PI). Fig.4.21. represents the cell cycle pattern of CF-DA-PEG treated and untreated cells. The A549 cells which are treated with different concentrations of CF-DA-PEG (5, 10 and 12 μg/mL) shows a typical DNA pattern that represented as sub-G1, G1, S, and G2/M phases of the cell cycle (Fig.4.21.). Due to treatment with 5 &10 μg/mL of CF-DA-PEG, the cells showed about 71% & 74 % population in G1 phase and the values are higher than the control set i.e. about 68 %. At same concentration of doses, we have observed that the percentages of cells are decreased in G2/M phase (16%&15%) compared to control (20 %) Where at higher dose (12 µg/mL), the percentages of sub-G1 phase (apoptotic cells) were significantly increased (9%) compare with control (1%) (Fig.4.21.). So, our particle CF-DA-PEG induces G1-phase cell cycle arrest at low concentration (5&10 µg/mL) and cell death at concentration of 12µg/mL.

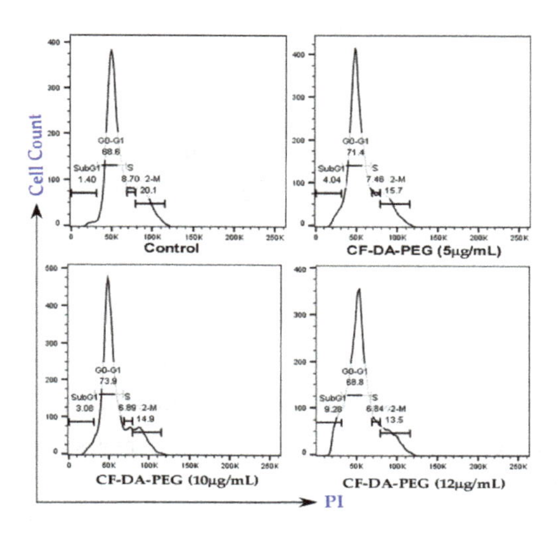

Fig.4.21. Flow cytometric analysis of cell cycle phase distribution of nuclear DNA in A549 cells after 24 h incubation with CF-DA-PEG. The histogram represents the DNA content (PI fluorescence along x-axis) versus cell count (along y-axis).

4.3.3.2.9. Western blot analysis in treated A549 cells:

There is another pathway which can also induce cell death in coupled with ROS generation. It is known that, p53 is a tumor suppressor protein which has activity to stop the formation of tumors and in human cancer cells the tumor suppressor gene remains inactive due to loss of p53 functions [74,75]. Normally when cells go through stress condition, the tumor suppressor protein p53 becomes activated and causes cell cycle arrest or apoptosis [76]. Hence, we have observed the p53 protein expression in untreated and treated cell by tagging the p53 with FITC conjugated anti- p53 antibody (Fig.4.22.(a)) and also by western blotting (Fig.4.22.(b)). In Fig.4.22.(a), we have seen that the p53expression is very low in control cells and the expression becomes increased after treatment with CF-DA-PEG in a dose dependent manner. Here, DAPI has been used as nucleus staining dye. The Fig.4.22.(a) data is re-established by the western blotting data (Fig.4.22.(b)) which also shows the elevation of p53 expression in treated cell than control. Therefore, CF-DA-PEG is successfully able to induce the p53 mediated apoptosis pathway in A549 cells. There are other sets of proteins which also play an important role in apoptosis pathway some proteins of them such as Bax, caspases promote apoptosis, while some members of Bcl2 family of proteins inhibit apoptosis. Generally, during apoptosis process, the cytochrome c gets released from mitochondria into the cytosol due to action of the Bax protein. After releasing, the cytochrome c binds with pro-caspase-9 in the cytosol and creates a protein complex known as apoptosome which cleaves the pro-caspase to its active form of caspase-9. The active caspase-9 then cleaves and activates pro-caspase into the active caspase-3. Bcl2 family of proteins inhibit the releasing of cytochrome c from mitochondria. We have observed the expression of Bax, Bcl2, caspase -9 and caspase-3 in treated and untreated cells by western blotting experiment (Fig.4.22.(b)) and noticed that the expression of apoptogenic protein Bax is increased and the anti-apototic protein Bcl2 expression is decreased after treatment with CF-DA-PEG compare to control. The pro-caspase-9 and pro-caspase-3 are also cleaved here and formed active caspase-9 and caspase-3 in treated cells. Hence, it is confirmed that CF-DA-PEG promotes the apoptotic process and causes cellular death.

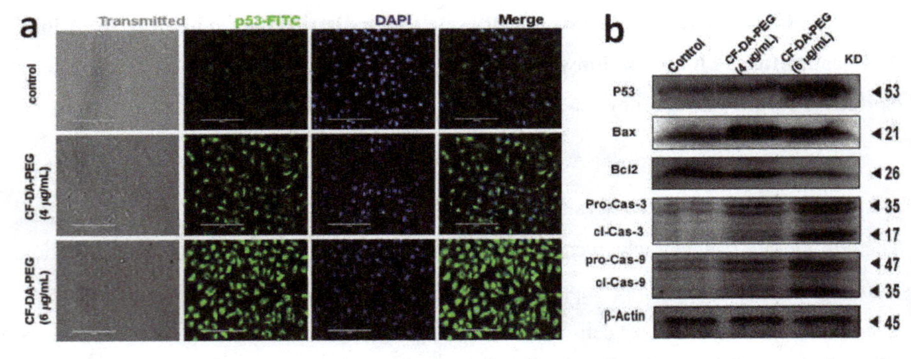

Fig.4.22. (a) Immunofluorescence images of A549 cells showing increased expression of p53 in the cytosol upon treatment with CF-DA-PEG in a dose dependent manner. The cells were stained with FITC conjugated anti- p53 (green) and nuclei with DAPI (blue). (b) Western blot analysis of p53, Bax, Bcl-2, caspase-3 and caspase-9 in untreated and CF-DA-PEG treated A549 cells. β-actin was used as control.

Now, we have checked the m-RNA expression of p53, Bax and Bcl2 by performing real-time PCR are shown in Fig.4.23. Here, we have observed that the p53 and Bax expression become increased in CF-DA-PEG treated cells while the Bcl2 level is decreased which supports the western blot data.

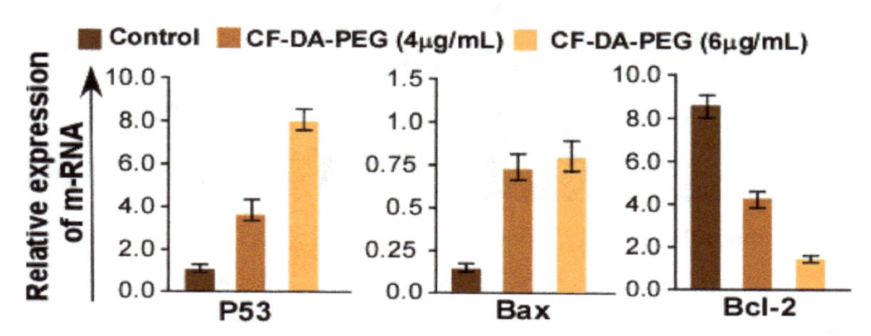

Fig.4.23. Graphical presentation of m-RNA expression of p53, Bax, Bcl-2 in untreated and CF-DA-PEG treated A549 cells.

4.3.3.2.10. Effects of CF-DA-PEG on cytochrome c expression:

Next, we have checked the expression of cytochrome c in mitochondria and cytosol. It is known that the cytochrome c is released from mitochondria to cytosol when a cell goes through apoptotic pathway. So, the elevated level of cytosolic cytochrome c confirms that the cells are in apoptosis path-way. In our experiment,the cytochrome c expression has been checked by fluorescence imaging after tagging the cytochrome c with FITC conjugated anti-cytochrome c antibody (Fig.4.24.(a)) and also by western blotting (Fig.4.24.(b)). In both cases, the cytosolic cytochrome c level is increased in treated cells in a dose dependent manner while in control cells the expression is very low. On other side, the mitochondrial cytochrome c level is decreased after treatment have noticed in Fig.4.24.(b) but in case of control the result is opposite. In Fig.4.24.(a), both treated and untreated cell nucleus was stained with DAPI. This data again confirms that CF-DA-PEG promotes the A549 cell apoptosis and causes cancer cell death.

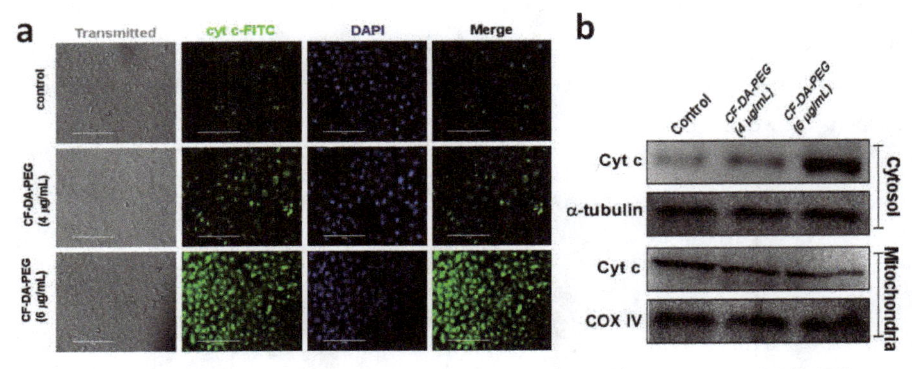

Fig.4.24. (a) Immunofluorescence images of cytochrome c- FITC (green) and DAPI (blue) stained control and treated A549 cells. (b) Western blot analysis of Cyt C expression level in mitochondria and cytosol of A549 cells after treatment with CF-DA-PEG. Here, α- tubulin and COX IV was kept as control for cytosol and mitochondria respectively.

4.3.3.2.11. Anti-migratory effect of CF-DA-PEG on A549 cells:

Like every cancer cells, A549 cells also shows metastatic behaviour which can be investigated by migratory property of the cells. By using both in-vitro bidirectional wound healing assay (Fig.4.25.(a)) & transwell migration assay (Fig.4.26.(a)), we have checked the anti-migratory property of treated and untreated cells. In Fig.4.25.(a), the dose dependent inhibition of cell migration have been noticed after treating the cells with CF-DA-PEG compared to control set. The images which were taken at 0 h after creating bidirectional wound, shows a big wound but after 24 h incubation almost all cells are migrated to two pole of wound in case of control cells. The CF-DA-PEG is efficient to stop cell migration completely at very low dose (2.5μg/mL). Fig.4.25.(b) showing the graphical representation of percentage of cell migration in control and treated cells after 24 h of incubation.

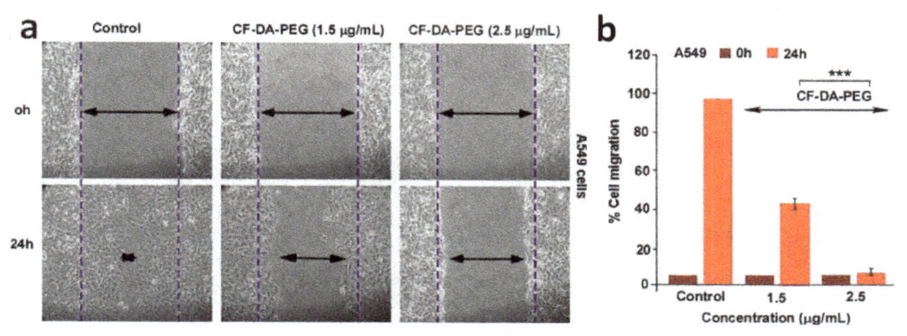

Fig.4.25. (a) Bidirectional wound healing assay determining the migration rate of A549 cells upon treatment with CF-DA-PEG (1.5 & 2.5 µg/mL) at 0 h and 24 h. (b) Graphical representation of bidirectional wound healing assay showing percentage of A549 cell migration at 0 h and 24 h on treatment with CF-DA-PEG. Here, the percentage of control cell migration at 24 h has been marked as 100% and the comparison have been made with this control.

The transwell migration assay (Fig.4.26.(a)) also supports the anti-migratory property of CF-DA-PEG on A549 cells. In this assay, the A549 cells were treated with specific doses of CF-DA-PEG and have been noticed that in case of treated cells lesser number of cells were migrated after 24 h of incubation where a significant high number of cells were migrated in case of control. After treatment, the cell migration is decreased in a dose dependent manner which supports the results of wound healing assay. The result of transwell migration assay has represented in graph where percentage of cell migration is plotted against control and different doses of CF-DA-PEG treated A549 cells (Fig.4.26.(b)). Hence, it has been concluded that the CF-DA-PEG was capable to retard the cell migration on A549 cell line even at very low dose.

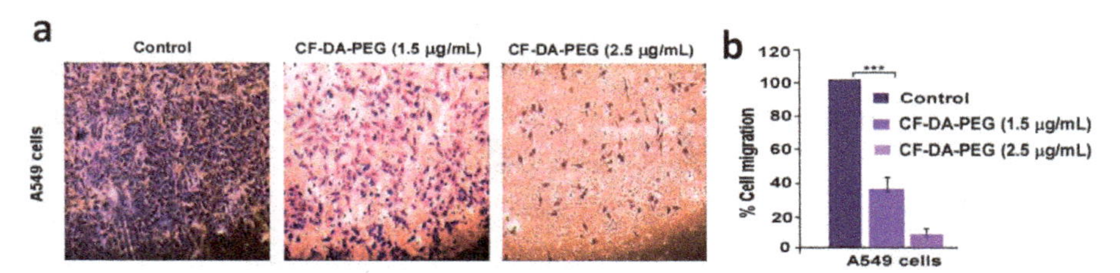

Fig.4.26. (a) Transwell migration assay images of A549 cells showing migration of cells upon treatment with different doses of CF-DA-PEG (1.5 & 2.5 µg/mL). (b) Graphical

representation of transwell migration assay presenting the percentage of cell migration after CF-DA-PEG treatment at same doses (1.5 & 2.5 μg/mL).

4.3.3.2.12. Effect of CF-DA-PEG on cell morphology:

Then we have checked the changes of cell morphology after treatment with different doses of CF-DA-PEG by using SEM imaging, which is shown in Fig.4.27. Here, we have seen that numerous protrusions are present on surface of the untreated cells where these protrusions significantly decrease gradually with increase of dose of CF-DA-PEG, which decrease the cell-cell contact too and upon treatment with 3μg/mL of CF-DA-PEG, the shrinking of cell membrane occurs and finally the cells are destructed.

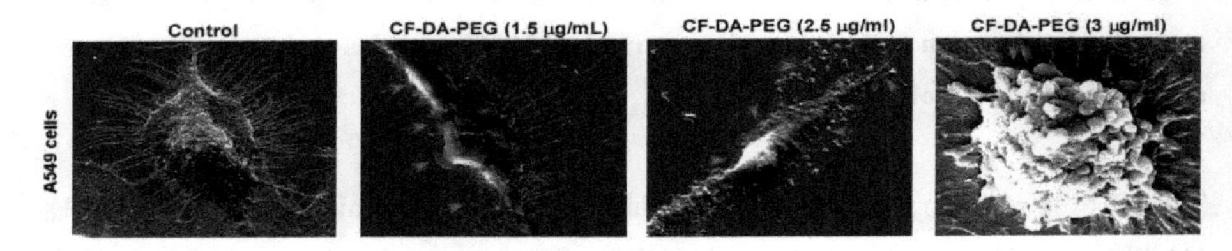

Fig.4.27. SEM images of A549 cells showing changes in cell morphology of A549 cells upon treatment with different doses of CF-DA-PEG compared to control. Arrows indicate the destruction of cell-cell junction and reduction in lammelipodia formation.

Inside the cells, the cytoskeleton plays an important role in some cellular activities such as cell signalling, cell division, motility and organelle transport etc. The cytoskeleton is formed by polymerization of actin filament which can label with a dye Phalloidin in fixed and permeabilized cells. We have noticed the effect of CF-DA-PEG on cytoskeleton formation by using phalloidin staining (Fig.4.28.) The CF-DA-PEG causes the destruction of cytoskeleton structure by inhibiting the actin polimerization where in control cells, the actin cytoskeleton is present in intact form which is shown in magnified image of Fig.4.28. Here, the nucleus of treated and control cells were stained with DAPI. So, CF-DA-PEG was able to stop the cytoskeleton formation and causes cell membrane destruction.

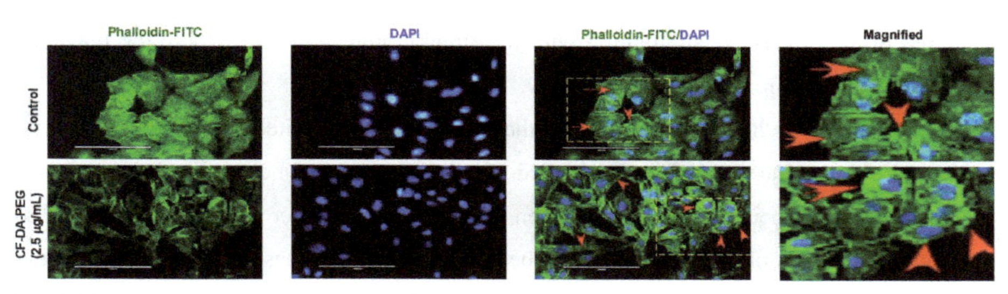

Fig.4.28. Immunofluorescence images of Phalloidin-FITCstained control and CF-DA-PEG treated A549 cells showing the actin cytoskeleton of cells. Cells were counter stained with DAPI. The arrows indicate the intact and disrupted actin cytoskeleton in the control and treated cells respectively.

4.4. Conclusion:

In conclusion, we have successfully synthesized the cube shaped magnetite nanoparticles by simple chemical precipitation method using surfactant. Structural characterization of this particle confirms that the particles are pure phase of Fe_3O_4 MNPs. The magnetic measurements give assurance for application of this MNP as drug delivery by heat triggered way. On other side, pH responsive drug release is also possible by this magnetic particle. The combined effect of pH and temperature shows better controlled drug release from Fe_3O_4 MNPs in a non-invasive way which gives new information in our study for better treatment of cancer and other diseases. Higher rate of drug release from this particles have been noticed at 43°C and pH5. The Fe_3O_4 MNPs shows the magnetic property under AC magnetic field and this particle behaves ferromagnetically. The magnetite particles are able to produce heat by rising the temperature upto 43°C in the solution at 600 Hz applied AC frequency. So, by applying AC magnetic field from outside the body, the controlled drug release can be achieved. The efficiency of drug loading with this particle is also much better compare to other published results. Therefore, the Fe_3O_4 MNPs offer a new way in the area of drug delivery and drug release.

Then, we have successfully prepared the spherical CFMNPs with two different sizes i.e. 250nm and 350 nm. Drug release studies indicated that the particles have a very good efficiency to AC magnetic field controlled drug release. Under AC magnetic field, the particles become agitated because of hysteresis loss that loose the drug molecules bound to particles and causes drug release. The bigger size of particles shows better DOX release compared to small size particles. Hence, the particles will be used for AC magnetic field triggered drug delivery by stimulating with the suitable magnetic field where the magnetic field can be applied from outside of the body.

In the next work, we have successfully synthesized the cobalt ferrite (CF) nanoaprticles and loaded with dopamine (DA). The DA loaded CF particles when functionalized with PEG, it provides good therapeutic option against non small cell lung cancer (NSCLC). The UV-Vis absorption spectroscopic data confirms that better DA has been released at low pH (pH5) and at high temperature (43°C). We have found that the CF-DA-PEG decreases the A549 cell viability and simultaneously induced ROS inside the cells. The CF-DA-PEG also helps in inducing of apoptosis and causes mitochondrial dysfunction. In addition, the CF-DA-PEG also restricts the migration of A549 cells and after treatment, the cell morphology has been changed and cell membrane shrinking takes place at certain dose. Hence, by using this PEG functionalized DA loaded particles, DA has been successfully delivered to the cancer site, which causes cellular death by inducing apoptosis and retarding migration which may be broadly applicable for treatment of non small cell lung cancer.

Bibliography:

[1] Yin, N.Q., Wu, P., Yang, T.H., Wang, M., 2017. Preparation and study of a mesoporous silica-coated Fe_3O_4 photothermal nanoprobe. RSC Adv., 7, 9123-9129.

[2] Feng, C., Pang, X., He, Y., Chen, Y., Zhang, G., Lin, Z., 2015. A versatile strategy for uniform hybrid nanoparticles and nanocapsules. Polym. Chem., 6, 5195197.

[3] Chen, Y., Wang, Z., Harn, Y.W., Pan, S., Li, Z., Lin, S., Peng, J., Zhang, G., Lin, Z., 2019. Resolving Optical and Catalytic Activities in Thermoresponsive Nanoparticles by Permanent Ligation with Temperature-Sensitive Polymers. Angew. Chemie Int. Ed., 58 (34), 11910-11917.

[4] Li, X., Iocozzia, J., Chen, Y., Zhao, S., Cui, X., Wang, W., Yu, H., Lin, S., Lin, Z., 2018. From precision synthesis of block copolymers to properties and applications of nanoparticles. Angew. Chemie Int. Ed., 57 (8), 2046-2070.

[5] Chen, Y., Yang, D., Yoon, Y.J., Pang, X., Wang, Z., Jung, J., He, Y., Harn, Y.W., He, M., Zhang, S., Zhang, G., Lin, Z., 2017. Hairy Uniform Permanently Ligated Hollow Nanoparticles with Precise Dimension Control and Tunable Optical Properties. J. Am. Chem. Soc., 139 (37), 12956-12967.

[6] Chen, Y., Wang, Z., He, Y., Yoon, Y.J., Jung, J., Zhang, G., Lin, Z., 2018. Light-enabled reversible self-assembly and tunable optical properties of stable hairy nanoparticles. Proc. Natl. Acad. Sci., 115 (7), E1391-E1400.

[7] Dey, C., Ghosh, A., Ahir, M., Ghosh, A., Goswami, M.M., 2018. Improvement of Anticancer Drug Release by Cobalt Ferrite Magnetic Nanoparticles through Combined pH and Temperature Responsive Technique. ChemPhysChem, 19, 2872-2878.

[8] Dey, C., Baishya, K., Ghosh, A., Goswami, M.M., Ghosh, A., Mandal, K., 2017. Improvement of drug delivery by hyperthermia treatment using magnetic cubic cobalt ferrite nanoparticles. J. Magn. Magn. Mater, 427, 168-174.

[9] Kyeong, S., Jeong, C., Kang, H., Cho, H.-J., Park, S.-J., Yang, J.-K., Kim, S., Kim, H.-M., Jun, B.-H., Lee, Y.-S., 2015. Double-layer magnetic nanoparticle-embedded silica particles for efficient bio-separation. PLoS One, 10, e0143727.

[10] Qiao, R., Yang, C., Gao, M., 2009. Superparamagnetic iron oxide nanoparticles: from preparations to in vivo MRI applications. J. Mater. Chem., 19, 6274-6293.

[11] Baselt, D.R., Lee, G.U., Natesan, M., Metzger, S.W., Sheehan, P.E., Colton, R.J., 1998. A biosensor based on magnetoresistance technology. Biosens. Bioelectron, 13 (7-8), 731-739.

[12] Corr, S.A., Rakovich, Y.P., Gun'ko, Y.K., 2008. Multifunctional Magnetic-fluorescent Nanocomposites for Biomedical Applications. Nanoscale Res. Lett., 3 (3), 87-104.

[13] Du, J.Z., Du, X.J., Mao, C.Q., Wang, J., 2011. Tailor-Made Dual pH-Sensitive PolymerDoxorubicin Nanoparticlesfor Efficient Anticancer Drug Delivery. J. Am. Chem. Soc.,133, 17560-17563.

[14] Goswami, M.M., 2016. Synthesis of Micelles Guided Magnetite (Fe3O4) Hollow Spheres and Their Application for AC Magnetic Field Responsive Drug Release. Nature publishing group, 6, 35721.

[15] Farokhzad, O.C., Langer, R., 2009. Impact of Nanotechnology on DrugDelivery. ACS Nano, 3, 16-20.

[16] Peer, D., Karp, J.M., Hong, S., Farokhzad, O.C., Margalit, R., Langer, R., 2007. Nanocarriers as an emerging platform for cancer therapy. Nature Nanotechnology, 2, 751-760.

[17] Torchilin, V.P., 2007. Micellar nanocarriers: pharmaceutical perspectives. Pharm. Res., 24(1), 1-16.

[18] Ulbrich, K., Subr, V., 2004. Polymeric anticancer drugs with pH-controlled activation. Adv. Drug Delivery Rev., 56 (7),1023-1050.

[19] Etrych, T., Jelinkova, M., Rihova, B., Ulbrich, K., 2001. New HPMA copolymers containing doxorubicin bound via pH-sensitive linkage: synthesis and preliminary in vitro and in vivo biological properties. J. Controlled Release, 73 (1), 89-102.

[20] Siegel, R.A., Firestone, B.A., 1988. pH-dependent equilibrium swelling properties of hydrophobic polyelectrolyte copolymer gels. Macromolecules, 21, 3254-3259.

[21] Yoshida, R., Uchida, K., Kaneko, Y., Sakai, K., Kikuchi, A., Sakurai, Y., Okano, T., 1995. Comb-type grafted hydrogels with rapid deswelling response to temperature changes. Nature, 374, 240-242.

[22] Sahoo, B., Devi, K.S.P., Banerjee, R., Maiti, T.K., Pramanik, P., Dhara, D., 2013. Thermal and pH responsive polymer-tethered multifunctional magnetic nanoparticles for targeted delivery of anticancer drug. ACS Appl. Mater. Interfaces, 5 (9), 3884-3893.

[23] Lee, S., Cha, E.-J., Park, K., Lee, S.-Y., Hong, J.-K., Sun, I.-C., Kim, S.Y., Choi, K., Kwon, I.C., Kim, K., Ahn, C.-H., 2008. A near-infrared-fluorescence-quenched gold-nanoparticle imaging probe for in vivo drug screening and protease activity determination. Angew. Chemie Int. Ed., 47 (15), 2804-2807.

[24] Gao, W., Chan, J.M., Farokhzad, O.C., 2010. pH-Responsive nanoparticles for drug delivery. Mol. Pharm., 7 (6), 1913-1920.

[25] Karimi, Z., Abbasi, S., Shokrollahi, H., Yousefi, G., Fahham, M., Karimi, L., Firuzi, O., 2017. Pegylated and amphiphilic Chitosan coated manganese ferrite nanoparticles for pH-sensitive delivery of methotrexate: Synthesis and characterization. Mater Sci Eng C Mater Biol Appl., 71, 504-511.

[26] Reddy, A.B., Manjula, B., Jayaramudu, T., Sadiku, E.R., Babu, P.A., Selvam, S.P., 2016. 5-Fluorouracil Loaded Chitosan–PVA/Na^+MMT Nanocomposite Films for Drug Release and Antimicrobial Activity. Nano-Micro Letters, 8(3), 260–269.

[27] Majewski, A.P., Schallon, A., Jerome, V., Freitag, R., Muller, A.H.E., Schmalz, H., 2012. Dual-Responsive Magnetic Core–Shell Nanoparticles for Nonviral Gene Delivery and Cell Separation. Biomacromolecules, 13 (3), 857-866.

[28] Chan, N., Laprise-Pelletier, M., Chevallier, P., Bianchi, A., Fortin, M.A., Oh, J.K., 2014. Multidentate Block-Copolymer-Stabilized Ultrasmall Superparamagnetic Iron Oxide Nanoparticles with Enhanced Colloidal Stability for Magnetic Resonance Imaging. Biomacromolecules, 15 (6), 2146-2156.

[29] De D.; Goswami M.M., 2016. Shape induced acid responsive heat triggered highly facilitated drug release by cube shaped magnetite nanoparticles. Biomicrofluidics, 10, 064112.

[30] Scharovsky, O.G., Matar, P., Rozados, V.R., et al., 2012. Immunomodulation and antiangiogenesis in cancer therapy. From basic to clinical research. Medicina (B Aires),72(1), 47–57.

[31] Nowak-Sliwinska, P., Griffioen, A.W., 2017. Angiogenesis inhibitors in combinatorial approaches. Angiogenesis, 20(2), 183–184.

[32] Mody, V.V., Cox, A., Shah, S., Singh, A., Bevins, W., Parihar, H., 2014. Magnetic nanoparticle drug delivery systems for targeting tumor. Appl. Nanosci., 4, 385–392.

[33] Maaz, K., Usman, M., Karim, S., Mumtaz, A., Hasanain, S.K., Bertino, M.F., 2009. Magnetic response of core-shell cobalt ferrite nanoparticles at low temperature. J. Appl. Phys., 105, 113917.

[34] Sarkar, D., Mandal, M., Mandal, K., 2014. Detail Study on ac–dc Magnetic and Dyeabsorption Properties of Fe3O4 Hollow Spheres for Biological and Industrial Application. J. Nanosci. Nanotechnol., 14, 2307–2316.

[35] Reddy, L.H., Arias, J.L., Nicolas, J., Couvreur, P., 2012. Magnetic Nanoparticles: Design and Characterization, Toxicity and Biocompatibility, Pharmaceutical and Biomedical Applications. Chem. Rev., 112, 5818–5878.

[36] Bacri, J-C., Silva, de Fatima Da Silva, M., Perzynski, R., Pons, J-N., Roger, J., Sabolovic', D., Halbreich, A., 1997. Use of Magnetic Nanoparticles for Thermolysis of Cells in a Ferrofluid. Scientific and Clinical Applications of Magnetic Carriers, 597–606.

[37] Iv, M., Telischak, N., Feng, D., Holdsworth, S.J., Yeom, K.W., Daldrup-Link, H.E., 2015. Clinical applications of iron oxide nanoparticles for magnetic resonance imaging of brain tumors. Nanomedicine, 10, 993–1018.

[38] Liu, T.Y., Hub, S.H., Liu, D.M., Chen, S.Y., Chen, I.W., 2009. Biomedical nanoparticle carriers with combined thermal and magnetic responses. Nano Today, 4, 52–65.

[39] Thiesen B., Jordan, A., 2008. Clinical applications of magnetic nanoparticles for hyperthermia. Int. J. Hyperthermia, 24 (6), 467–474.

[40] Huang, H., Delikanli, S., Zeng, H., Ferkey, D.M., Pralle, A., 2010. Remote control of ion channels and neurons through magnetic-field heating of nanoparticles. Nat. Nanotechnol., 5 (8), 602–606.

[41] Rosensweig, R.E., 2002. Heating Magnetic Fluid with Alternating Magnetic Field. J. Magn. Magn. Mater, 252 (1-3), 370–374.

[42] Decuzzi, P., Godin, B., Tanaka, T., Lee, S.Y., Chiappini, C., Liu, X., Ferrari, M., 2010. Size and shape effects in the biodistribution of intravascularly injected particles. J. Controlled Release, 141, 320–327.

[43] Shah, S., Liu, Y., Hu, W., Gao, J., 2011. Modeling Particle Shape-Dependent Dynamics in Nanomedicine. J. Nanosci. Nanotechnol., 11 (2), 919-928.

[44] Gratton, S.E.A., Ropp, P.A., Pohlhaus, P.D., Luft, J.C., Madden, V.J., Napier, M.E., DeSimone, J.M., 2008. The effect of particle design on cellular internalization pathways. Proc. Natl.Acad. Sci. U.S.A., 105 (33), 11613–11618.

[45] Gustafson, H.H., Casper, D.H., Grainger, D.W., Ghandehari, H., 2015. Nanoparticle Uptake: The Phagocyte Problem. Nano today, 10, 487-510.

[46] Blanco, E., Shen, H., Ferrari, M., 2015. Principles of nanoparticle design for overcoming biological barriers to drug delivery. Nature biotechnology, 33, 941-951.

[47] Zhao, N., Wu, S., He, C., Wang, Z., Shi, C., Liu, E., Li, J., 2013. One-pot synthesis of uniform Fe3O4 nanocrystals encapsulated in interconnected carbon nanospheres for superior lithium storage capability. Carbon, 57, 130–138.

[48] Cornell, R.M., Schwertmann, U., 1996. The Iron Oxides. Structure, Properties, Reactions, Occurrence and Uses. (Wiley-VCH, Weinheim, Germany).

[49] Hergt, R., Dutz, S., Muller, R. Zeisberger, M., 2006. Magnetic particle hyperthermia: nanoparticle magnetism and materials development for cancer therapy. J. Phys.: Condens. Matter, 18, S2919–S2934.

[50] Singh, N., Karambelkar, A., Gu, L., Lin, K., Miller, J.S., Chen, C.S., Sailor, M.J., Bhatia, S.N., 2011. Bioresponsive mesoporous silica nanoparticles for triggered drug release. J. Am. Chem. Soc., 133 (49), 19582–19585.

[51] Parrott, M.C., Finniss, M., Luft, J.C., Pandya, A., Gullapalli, A., Napier, M.E., DeSimone, J.M., 2012. Incorporation and Controlled Release of Silyl Ether Prodrugs from PRINT Nanoparticles. J. Am. Chem. Soc.,134 (18), 7978–7982.

[52] Lee, E.S., Gao, Z., Bae, Y.H., 2008. Recent progress in tumor pH targeting nanotechnology. J. Controlled Release, 132 (3), 164–170.

[53] Su, J., Chen, F., Cryns, V.L., Messersmith, P.B., 2011. Catechol Polymers for pH-Responsive, Targeted Drug Delivery to Cancer Cells. J. Am. Chem. Soc. 133, 11850–11853.

[54] Meng, H., Min, X., Xia, T., Zhao, Y.L., Tamanoi, F., Stoddart, J.F., Zink, J.I., Nel, A.E., 2010. Autonomous in Vitro Anticancer Drug Release from Mesoporous Silica Nanoparticles by pH-Sensitive Nanovalves. J. Am. Chem. Soc., 132,12690–12697.

[55] Zhang, J., Yuan, Z.-F., Wang, Y., Chen, W.-H., Luo, G.-F., Cheng, S.-X., Zhuo, R.-X., Zhang, X.-Z., 2013. Multifunctional Envelope-Type Mesoporous Silica Nanoparticles for Tumor-Triggered Targeting Drug Delivery. J. Am. Chem. Soc.,135, 5068–5073.

[56] Lai, J., Shah, B.P., Garfunkel, E., Lee, K.-B., 2015. Real-Time Monitoring of ATP-Responsive Drug Release Using Mesoporous-Silica-Coated Multicolor Upconversion Nanoparticles. ACS Nano, 9 (5), 5234–5245.

[57] Chaterjee, K., Jianqing, Z., Norman, H., Karliner, J.S., 2010. Doxorubicin Cardiomyopathy. Cardiology, 115 (2), 155–162.

[58] Kaczmarek, A., Brinkman, B.M., Heyndrickx, L., Vandenabeele, P., Krysko, D.V. J., 2012. Severity of doxorubicin-induced small intestinal mucositis is regulated by the TLR-2 and TLR-9 pathways. J. Pathol., 226 (4), 598–608.

[59] Lopez, T., Bata-Garcia, J.L., Esquivel, D., et al., 2010. Treatment of Parkinson's disease: nanostructured sol-gel silica-dopamine reservoirs for controlled drug release in the central nervous system. Int J Nanomedicine, 6, 19–31.

[60] Shameli, K., Ahmad, M.B., Jazayeri, S.D., Sedaghat, S., Shabanzadeh, P., Jahangirian, H., Mahdavi, M., Abdollahi, Y., 2012. Synthesis and Characterization of Polyethylene Glycol Mediated Silver Nanoparticles by the Green Method. Int. J. Mol. Sci., 13, 6639-6650.

[61] Peters, E.B., Christoforou, N., Leong, K.W., Truskey, G.A., West, J.L., 2016. Poly(ethylene glycol) Hydrogel Scaffolds Containing Cell-Adhesive and Protease-Sensitive Peptides Support Microvessel Formation by Endothelial Progenitor Cells. Cell Mol Bioeng., 9(1), 38–54.

[62] Sarkar, C., Chakroborty, D., Dasgupta, P.S., Basu, S., 2015. Dopamine is a safe antiangiogenic drug which can also prevent 5-fluorouracil induced neutropenia. Int. J. Cancer, 137(3), 744-749.

[63] Sivandzade, F., Bhalerao, A., Cucullo, L., 2019. Analysis of the Mitochondrial Membrane Potential Using the Cationic JC-1 Dye as a Sensitive Fluorescent Probe. Bio-protocol, 9(1), e3128.

[64] Basu, A., Upadhyay, P., Ghosh, A., Chattopadhyay, D., Adhikary, A., 2019. Folic-Acid-Adorned PEGylated Graphene Oxide Interferes with the Cell Migration of Triple Negative Breast Cancer Cell Line, MDAMB-231 by Targeting miR-21/PTEN Axis through NFκB. ACS Biomater. Sci. Eng., 5, 373−389.

[65] Delcambre, S., Nonnenmacher, Y., Hiller, K., 2016. Mitochondrial Mechanisms of Degeneration and Repair in Parkinson's Disease. Springer, 25-47.

[66] Sun Y.; Pham A.N.; Hare D.J.; Waite T.D., 2018. Kinetic Modeling of pH-Dependent Oxidation of Dopamine by Iron and Its Relevance to Parkinson's Disease. Front. Neurosci.

[67] Wang, J., Luo, B., Wen, S., 2017. Inhibition of cancer growth in vitro and in vivo by a novel ROS-modulating agent with ability to eliminate stem-like cancer cells. Cell Death & Disease, 8(6), e2887.

[68] Gherghi, I.C., Girousi, S.T., Voulgaropoulos, A., et al., 2003. Study of interactions between DNA-ethidium bromide (EB) and DNA-acridine orange (AO), in solution, using hanging mercury drop electrode (HMDE). Talanta, 61,103-112.

[69] Leite, M., Quinta-Costa, M., Leite, P.S., et al., 1999. Critical evaluation of techniques to detect and measure cell death-study in a model of UV radiation of the leukaemic cell line HL60. Anal cell Pathol.,19,139-151.

[70] Baskic, D., Popovic, S., Ristic, P., et al., 2006. Analysis of cycloheximide- induced apoptosis in human leukocytes: fluorescence microscopy using annexin V/propidium iodide versus acridin orange/ethidium bromide. Cell Biol Int., 30, 924-932.

[71] Liu, K., Liu, P.C., Liu, R., Wu, X., 2015. Dual AO/EB staining to detect apoptosis in osteosarcoma cells compared with flow cytometry. Med Sci Monit Basic Res., 21,15–20.

[72] Hingorani R.; Deng J.; Elia J.; McIntyre C.; Mittar, D., 2011. Detection of Apoptosis Usingthe BD Annexin V FITC Assayon the BD FACSVerse™ System. BD Biosciences.

[73] Murad, H., Hawat, M., Ekhtiar, A., et al., 2016. Induction of G1-phase cell cycle arrest and apoptosis pathway in MDA-MB-231 human breast cancer cells by sulfated polysaccharide extracted from Laurencia papillosa. Cancer Cell Int.,16, 39.

[74] Hussain, S.P., Harris, C.C., 1998. Molecular Epidemiology of Human Cancer. Recent Results Cancer Res.,154, 22-36.

[75] Beroud, C., Soussi, T., 2003. The UMD-p53 database: New mutations and analysis tools. Human Mutation, 21(3), 176-181.

[76] Haupt, S., Berger, M., Goldberg, Z., Haupt, Y., 2003. Apoptosis - the p53 network. Journal of Cell Science, 116, 4077-4085.

Chapter 5 — Conclusion and scope for future work

This chapter describes overall conclusion of the details work which are described in this thesis and also the scope and possibilities for further work in this research area.

5.1. Epilogue:

This thesis describes the detail synthesis, characterization and surface modification procedures of different types of magnetic nanoparticles (Fe_3O_4 & $CoFe_2O_4$) and their broad applications in cancer treatments.

We have synthesized magnetic nanoparticles (MNPs) by chemical co-precipitation method and functionalized them with some organic molecules such as DNA, PEG etc. to increase their biocompatibility. After synthesis and functionalization, these MNPs were characterized in detail. Structural characterization was done by XRD and FTIR study where TEM has been used for morphological analysis. Magnetic measurements were carried out in VSM and SQUID. Binding affinity between nanoparticles and organic molecules has been checked by ITC. We have used these MNPs for different area of applications according to their properties. In our study, we have used the MNPs for hyperthermia applications and drug delivery to treat different types of cancer cells. Here, we have checked the effect of $CoFe_2O_4$ NP in the field of hyperthermia on triple negative breast cancer cells (TNBC) (MDAMB-231 cell lines). In this experiment, comparative studies have been performed between bare $CoFe_2O_4$ (CF) NP and DNA engineered $CoFe_2O_4$ (CF-DNA) NP and observed that more CF-DNA NPs has been internalized inside the MDAMB-231 cell lines compare to CF NPs. Due to greater entry of CF-DNA NP, it has been observed slight higher cytotoxic effect of DNA functionalized particles TNBC cells. Both the particles show biocompatibility on PBMC (peripheral blood mononuclear cell) at certain concentrations of treatment but at higher dose they may show some toxicity. Therefore, we have performed the hyperthermia applications where at very low dose of treatment, the cancer cell death occurs. Details hyperthermia applications were done on our lab made instrument and we have observed that the CF NP generates heat under AC magnetic field and the particles gain heat quickly under the AC field but it takes long time to cool down. In case of AC measurement, the power loss that measured from the AC hysteresis loops is sufficient to generate heat causing rise of temperature for hyperthermia treatment. Due to DNA binding, the magnetic property of CF NP has been changed which helps to tune the magnetic property for suitable heat generation as per required. The cell morphology has been checked under SEM where shrinkage of cell membrane has been observed due to hyperthermia applications.

We also investigate the effect of Fe_3O_4 and $CoFe_2O_4$ NPs on drug delivery at cancer cells. Different stimuli (temperature and pH) dependent drug delivery has been checked to improve the technique and to increase the efficiency of drug. Two drugs have been used as delivering agent such as doxorubicin and dopamine. In our study Fe_3O_4 shows good ferromagnetic property which helps release of greater amount of doxorubicin drugs to cancer cell regions triggering by hyperthermia therapy. To make the nanoparticles more biocompatible and well dispersed in solution, surface modifications of drug loaded $CoFe_2O_4$ has been done with PEG. PEG also increases the sustainability of drug loaded particles in dissolved condition. To check drug delivery in cancer cells, the PEG coated drug loaded particles (CF-DA-PEG) were passed through the detail cellular experiments where the particles showed toxic effects on A549 cells even at low dose and induces apoptosis by ROS generation inside the cells. Tagging the NPs with fluorescence marker confirms the better internalization of the nanoparticles by cancer cells. Our particles also cause mitochondrial dysfunction and restrict the migration. Therefore treatment with CF-DA-PEG successfully released drugs at non small cell lung cancer and causes cellular death.

5.2. Scope for future work:

In research area, ferrites NPs have already taken attention due to their wide range of applications. Our synthesized nanoparticles are ferromagnetic in character which show good magnetic property and also successfully can be used for hyperthermia, but if instrument with high frequency range can be availed, besides of ferromagnetic material superparamagnetic materials may also be utilized to generate heat and may be used in hyperthermia. Because our synthesized nanoparticles have shown their applicability both in drug delivery and hyperthermia triggered drug delivery, it will open up a new way for treatment of cancer or other diseases where similar type of therapy might be required. Here, we have synthesized the NPs with different size and shapes and checked their physiocochemical properties for using these in drug delivery and hyperthermia. So, in futute, one can get idea about the suitability of use of these particles. Functionalizations of the MNPs with organic molecules improve their application efficiency, biocompatibility, reduce the agglomeration and also increases the circulation time. For cellular applications, these parameters are necessary. We have imaged the cancer cells by tagging the bare and functionalized MNPs with fluorescent marker and a compertaive studies have been done. Here, we have used two drugs i.e. doxorubicin and dopamine for drug delivery. In future, the drug delivering property of other drugs can also be checked. Dopamine is a neurotransmitter that naturally produced in the body, so it has no any adverse effect on human body. We have first observed its anti-cancer effect on lung cancer cell lines. We have performed the detail in-vitro experiments on dopamine delivery by using magnetic nanoparticles and also find out the path by which the cell apoptosis occurs. In future, in-vivo experiments are necessary. We have synthesized all the nanoparticles by very simple method with very low cost and also loaded the drugs with them by easy technique. Therefore, by using these protocols, the treatment cost of cancer can be reduced. Our drug loaded particles causes the cancer cell death even at very low dose. Here, we have carried out our experiments on lung cancer cell line (A 549) and human mammary carcinoma cell line (MDA MB 231). We can check the effects of these particles on other cell-line also.

CPSIA information can be obtained
at www.ICGtesting.com
Printed in the USA
BVHW051232110423
662129BV00006B/485

9 787238 149046